Fundamentals and Recent Advances in Epitaxial Graphene on SiC

Fundamentals and Recent Advances in Epitaxial Graphene on SiC

Editors

Rositsa Yakimova
Ivan Shtepliuk

MDPI • Basel • Beijing • Wuhan • Barcelona • Belgrade • Manchester • Tokyo • Cluj • Tianjin

Editors
Rositsa Yakimova
Semiconductor Materials,
Department of Physics,
Chemistry and Biology-IFM
Linköping University
Sweden

Ivan Shtepliuk
Semiconductor Materials,
Department of Physics,
Chemistry and Biology-IFM
Linköping University
Sweden

Editorial Office
MDPI
St. Alban-Anlage 66
4052 Basel, Switzerland

This is a reprint of articles from the Special Issue published online in the open access journal *Applied Sciences* (ISSN 2076-3417) (available at: https://www.mdpi.com/journal/applsci/special_issues/epitaxial_graphene_on_sic).

For citation purposes, cite each article independently as indicated on the article page online and as indicated below:

LastName, A.A.; LastName, B.B.; LastName, C.C. Article Title. *Journal Name* **Year**, *Volume Number*, Page Range.

ISBN 978-3-0365-1179-5 (Hbk)
ISBN 978-3-0365-1178-8 (PDF)

Contents

About the Editors

Rositsa Yakimova

Rositsa Yakimova is a professor emerita in material science at the Department of Physics, Chemistry, and Biology, Linköping University. She is an internationally recognized expert in the field of semiconductor crystals and nanostructure growth of wide bandgap materials, e.g., SiC, AlN, ZnO, and gapless graphene. She has made substantial contributions to the development of the sublimation growth process. Her major efforts since the demonstration of graphene have been in the field of epitaxial graphene on SiC. Yakimova has pioneered a novel method for the fabrication of large areas of uniform graphene, which is patent-protected. Since 2008, she has led the research of graphene on SiC at Linköping University, and for a long time, she was a member of Graphene Flagship. In total, she has authored over 500 publications and has received 9000 citations. Prof. R. Yakimova is a cofounder and a member of the Board of Graphensic AB, a spin-off company from Linköping University.

Ivan Shtepliuk

Ivan Shtepliuk is a researcher at the Department of Physics, Chemistry, and Biology, Linköping University. He received his Ph.D. in Solid State Physics (2013) at the Institute for Problems in Materials Science (Kyiv, Ukraine). During the years 2013 to 2015, he worked as a senior researcher at the Institute for Problems in Materials Science, National Academy of Sciences of Ukraine (Kyiv, Ukraine), where he was engaged in the development of ZnO-based light-emitting diodes. He has since moved to Linköping University in 2014. His research focus has shifted to graphene-based materials, such as graphene quantum dots, epitaxial graphene on SiC, and metal-decorated graphene, which are applicable in various technological devices, including heavy metal sensors, gas sensors, and Schottky diodes. He is also an expert in theoretical modelling using density functional theory methods. He has authored more than 60 papers published in reputable peer-reviewed journals.

Editorial

Special Issue "Fundamentals and Recent Advances in Epitaxial Graphene on SiC"

Ivan Shtepliuk * and Rositsa Yakimova

Semiconductor Materials, Department of Physics, Chemistry and Biology-IFM, Linköping University, SE-58183 Linköping, Sweden; rositsa.yakimova@liu.se
* Correspondence: ivan.shtepliuk@liu.se

Abstract: The aim of this Special Issue is to provide a scientific platform for recognized experts in the field of epitaxial graphene on SiC to present their recent studies towards a deeper comprehension of growth mechanisms, property engineering and device processing. This Special Issue gives readers the possibility to gain new insights into the nature of buffer layer formation, control of electronic properties of graphene and usage of epitaxial graphene as a substrate for deposition of different substances, including metals and insulators. We believe that the papers published within the current Special Issue develop cumulative knowledge on matters related to device-quality epaxial graphene on SiC, bringing this material closer to realistic practical applications.

Keywords: epitaxial graphene; sublimation; SiC; buffer layer; electronic properties; material engineering; deposition

Citation: Shtepliuk, I.; Yakimova, R. Special Issue "Fundamentals and Recent Advances in Epitaxial Graphene on SiC". *Appl. Sci.* **2021**, *11*, 3381. https://doi.org/10.3390/app11083381

Received: 12 March 2021
Accepted: 30 March 2021
Published: 9 April 2021

Publisher's Note: MDPI stays neutral with regard to jurisdictional claims in published maps and institutional affiliations.

1. Introduction

For more than a decade, investigations of epitaxial graphene on SiC have gained special urgency in view of its possible applications in many fields, including metrology, electronics and sensorics. Further progress in the development of related technologies requires both rethinking of already existing knowledge and discovery of innovative solutions. This was the primary motivation for opening the call for papers within the Special Issue "Fundamentals and Recent Advances in Epitaxial Graphene on SiC".

In total, the Special Issue encompasses four research papers and three review papers. Two research works touch on crucial aspects of early stage of graphene growth, namely buffer layer formation [1], and graphene quality estimation [2]. Kaushik et al. [3] reported on a principal possibility to tune structural and electronic properties of epitaxial graphene through nitrogen ion implantation. Fundamental knowledge on both copper electrodeposition and atomic layer deposition of high-k insulators on epitaxial graphene/SiC is provided by the authors of [4,5]. These results suggest that epitaxial graphene is a stable support for metal and metal oxides, which is important in the context of metal contacts, gating, etc. Concomitantly, the interaction between epitaxial graphene and its environment, including metal contacts may limit, to some extent carrier transport in epitaxial graphene and therefore needs to be considered in detail. The role of such interaction has been a research subject of the review paper by Pradeepkumar et al. [6]. Finally, Wu et al. [7] critically reviewed recent advances in graphene twistronics and identified epitaxial graphene on SiC as the most promising platform for twistronics.

2. Critical Aspects of Epitaxial Graphene Growth: Recipes, Properties, and Quality

The quality of the buffer layer (also known as C-rich surface reconstruction of SiC and zero graphene layer) is identified as one of the most important factors determining the quality of the epitaxial graphene monolayer on SiC grown via Si sublimation approach. In other words, the fabrication of a large-area epitaxial graphene layer with high thickness uniformity requires pre-formation of high-quality continuous buffer layer on large areas.

1

Thus, an optimization of the growth regime with respect to the buffer layer formation during early stage graphenization process as well as a complete understanding of the growth mechanism are key ingredients to fabricate device-quality graphene. In fact, a successful graphenization process may occur only in a very narrow operational temperature/pressure window which imposes significant restrictions on the growth regime tunability. This makes the optimization of growth conditions quite a challenging task. Despite the large research efforts to tackle this task, it still requires more systematic consideration. In this regard, the critical study by the authors of [1] on optimizing the formation conditions of buffer layers through control of the graphite crucible temperature and varying the Ar gas pressure is a recent contribution to the process. It was revealed that the buffer layer coverage is strongly dependent on the temperature at which Ar gas is introduced, demonstrating a gradual decrease as the temperature increases. The mechanism behind this behavior has been discussed. In the same paper, the relationship between the growth temperature and electronic properties (carrier mobility, carrier density) of quasi-freestanding graphene monolayer and epitaxial graphene monolayer, respectively, were discussed. It was also illustrated that the conductivity type and free carrier density for graphene are extremely sensitive to ambient conditions which was observed by many researchers earlier. In line with this, the review paper by Pradeepkumar et al. [6] provides a more general picture of the effect of epitaxial graphene-ambient interaction on the carrier transport in SiC-supported epitaxial graphene. The authors highlight the adsorption of different molecules (O_2, H_2O, NO_2, H_2O_2, CO_2, NH_3, CO, NO, N_2O_4) as a main reason that underlie conductivity type flipping, transport properties fluctuations and carrier density saturation.

Apart from the unintentional doping of epitaxial graphene by environmental gases and other molecules, the electronic properties of epitaxial graphene on SiC can be modulated by the intentional incorporation of external dopants, as was demonstrated in another study [3]. Nitrogen ion implantation was proposed as an instrumental approach to stabilize the n-type conductivity in epitaxial graphene without serious structural damage. However, a balance between graphene quality and implantation dose must be reached. In that light, the mentioned paper dealt with finding the correlation between the fluence value and epitaxial graphene properties such as fragmentation degree, and defect density.

It is instructive that all mentioned works exploit Raman Spectroscopy to estimate the quality of epitaxial graphene. More specifically, the relationship between the intensities of $2D$ and G characteristic peaks is used to determine the number of graphene layers, while D/G amplitude ratio is employed to calculate the defect density. Although the graphene Raman spectroscopy is a mature field, it continues to evolve especially in the direction of signal processing (for example, peak fitting quality). In this regard, the work by Kunc and Rejhon [2] originally offers a Voigt line shape fitting approach for analysis of $2D$ peak line shape for epitaxial graphene, which includes both the inhomogeneous and homogeneous broadening. They also interpreted the physical nature of each term by ascribing the homogeneous broadening to intrinsic lifetime and inhomogeneous broadening to strain fluctuations, respectively.

3. Epitaxial Graphene as a Host for Material Deposition

Epitaxial graphene on SiC is of great interest because it not only has extraordinary intrinsic properties but also can be used as an atomically flat robust support for non-hybridized growth of different materials, especially metals and metal oxides. Such an integration may expand the functionality of epitaxial graphene and boost the development of innovative technologies in conceptually new fields, like catalysis, plasmonics, and spintronics. Thus, research efforts to contribute to this field and to enrich the existing knowledge capital are in high demand. In response to this demand, the group at Linköping University [4] launched a systematic study of metal electrodeposition on epitaxial graphene on SiC, choosing copper as a model metal at the first stage. This work sheds light on fundamental aspects of copper electrochemistry on epitaxial graphene and shows that copper electroreduction occurs via two subsequent single-electron transfer steps. The instantaneous nucleation mechanism was identified as a dominating mechanism during copper electrodeposition. The present results provide a deep understanding of the nature of copper–epitaxial graphene interaction, thereby facilitating the design of novel copper–graphene nanohybrid materials.

At the same time, Giannazzo et al. [5] in their work gave an overview of the recent results on the growth of high-k insulators on epitaxial graphene on SiC, focusing on atomic layer deposition of Al_2O_3 thin layers, which are important for fabrication of epitaxial graphene-based devices. It was argued that the monolayer epitaxial graphene uniformity is a key factor to achieve a homogeneous Al_2O_3 coverage via direct deposition, the latter has not been successful before in other studies. The role of different seeding layers and surface pre-functionalization in atomic layer deposition processes on epitaxial graphene is critically discussed. Finally, the authors explained the effect of pre-treatment and grown layers on the quality and electronic properties of epitaxial graphene. As was discussed in [6], this issue requires careful consideration, since the interaction between graphene and deposited layers may significantly affect the electron transport in graphene.

4. A New Look at Possible Applications of Epitaxial Graphene on SiC

Although the epitaxial graphene on SiC is nowadays reasoned to be utilized in electronics, quantum metrology, and gas/liquid sensing, the unique properties of this material make it promising for use in other non-conventional fields. Wu et al. [7] claim that epitaxial graphene on SiC could be regarded as an excellent platform for formation of twisted few-layer graphene with a magic twist angle that might be useful to control spin orders, ferromagnetism, and superconductivity. The authors have substantiated this claim by the fact that owing to its natural compatibility with the semiconductor technologies, epitaxial graphene-based device processing requires no intermediate graphene transfer steps and thus is more attractive from a technological point of view in comparison to transferred graphene. In this regard, there is a plenty of room for manipulation of the twist angle and for formation of twisted graphene layers on SiC with the desired angle through adjusting the sublimation growth conditions.

5. Concluding Remarks

The Guest Editors consider the current collection of papers as an important piece of the puzzle needed to boost both the more rational implementation of epitaxial graphene into traditional devices and the development of non-conventional innovative technologies. Furthermore, the new results reported in the frame of the Special Issue complement the existing knowledge on buffer layer formation, material preparation–property relationships, and growth mechanisms of different materials on epitaxial graphene. We believe that this information input will provide the driving force behind future experimental efforts to improve the epitaxial graphene quality and to design sophisticated devices exploiting epitaxial graphene as active and passive components.

Funding: This research received no external funding.

Institutional Review Board Statement: Not applicable.

Informed Consent Statement: Not applicable.

Acknowledgments: Guest editors are grateful to all authors, reviewers, and Applied Sciences editors for their important contributions to prepare the current Special Issue.

Conflicts of Interest: The authors declare no conflict of interest.

References

1. Stanishev, V.; Armakavicius, N.; Bouhafs, C.; Coletti, C.; Kühne, P.; Ivanov, I.G.; Zakharov, A.A.; Yakimova, R.; Darakchieva, V. Critical View on Buffer Layer Formation and Monolayer Graphene Properties in High-Temperature Sublimation. *Appl. Sci.* **2021**, *11*, 1891. [CrossRef]
2. Kunc, J.; Rejhon, M. Raman 2D Peak Line Shape in Epigraphene on SiC. *Appl. Sci.* **2020**, *10*, 2354. [CrossRef]

3. Kaushik, P.D.; Yazdi, G.R.; Lakshmi, G.B.V.S.; Greczynski, G.; Yakimova, R.; Syväjärvi, M. Structural Modifications in Epitaxial Graphene on SiC Following 10 keV Nitrogen Ion Implantation. *Appl. Sci.* **2020**, *10*, 4013. [CrossRef]
4. Shtepliuk, I.; Vagin, M.; Yakimova, R. Electrochemical Deposition of Copper on Epitaxial Graphene. *Appl. Sci.* **2020**, *10*, 1405. [CrossRef]
5. Giannazzo, F.; Schilirò, E.; Lo Nigro, R.; Roccaforte, F.; Yakimova, R. Atomic Layer Deposition of High-k Insulators on Epitaxial Graphene: A Review. *Appl. Sci.* **2020**, *10*, 2440. [CrossRef]
6. Pradeepkumar, A.; Gaskill, D.K.; Iacopi, F. Electronic and Transport Properties of Epitaxial Graphene on SiC and 3C-SiC/Si: A Review. *Appl. Sci.* **2020**, *10*, 4350. [CrossRef]
7. Wu, D.; Pan, Y.; Min, T. Twistronics in Graphene, from Transfer Assembly to Epitaxy. *Appl. Sci.* **2020**, *10*, 4690. [CrossRef]

Article

Critical View on Buffer Layer Formation and Monolayer Graphene Properties in High-Temperature Sublimation

Vallery Stanishev [1], Nerijus Armakavicius [1,2], Chamseddine Bouhafs [1], Camilla Coletti [3], Philipp Kühne [1,2], Ivan G. Ivanov [4], Alexei A. Zakharov [5], Rositsa Yakimova [4] and Vanya Darakchieva [1,2,*]

[1] Terahertz Materials Analysis Center, Department of Physics, Chemistry and Biology, IFM, Linköping University, 581 83 Linköping, Sweden; vallery.stanishev@liu.se (V.S.); nerijus.armakavicius@liu.se (N.A.); cebouhafs@gmail.com (C.B.); philipp.kuhne@liu.se (P.K.)
[2] Center for III-Nitride Technology C3NiT-Janzén, Department of Physics, Chemistry and Biology, IFM, Linköping University, 581 83 Linköping, Sweden
[3] Center for Nanotechnology Innovation @NEST, Istituto Italiano di Tecnologia, Piazza S. Silvestro, 12, 56127 Pisa PI, Italy; Camilla.Coletti@iit.it
[4] Department of Physics, Chemistry and Biology, IFM, Linköping University, 581 83 Linköping, Sweden; ivan.gueorguiev.ivanov@liu.se (I.G.I.); rositsa.yakimova@liu.se (R.Y.)
[5] MaxLab, Lund University, S-22100 Lund, Sweden; alexei.zakharov@maxiv.lu.se
* Correspondence: vanya.darakchieva@liu.se

Citation: Stanishev, V.; Armakavicius, N.; Bouhafs, C.; Coletti, C.; Kühne, P.; Ivanov, I.G.; Zakharov, A.A.; Yakimova, R.; Darakchieva, V. Critical View on Buffer Layer Formation and Monolayer Graphene Properties in High-Temperature Sublimation. *Appl. Sci.* **2021**, *11*, 1891. https://doi.org/10.3390/app11041891

Academic Editor: Vasili Perebeinos

Received: 31 December 2020
Accepted: 13 February 2021
Published: 21 February 2021

Publisher's Note: MDPI stays neutral with regard to jurisdictional claims in published maps and institutional affiliations.

Abstract: In this work we have critically reviewed the processes in high-temperature sublimation growth of graphene in Ar atmosphere using closed graphite crucible. Special focus is put on buffer layer formation and free charge carrier properties of monolayer graphene and quasi-freestanding monolayer graphene on 4H–SiC. We show that by introducing Ar at higher temperatures, T_{Ar}, one can shift the formation of the buffer layer to higher temperatures for both n-type and semi-insulating substrates. A scenario explaining the observed suppressed formation of buffer layer at higher T_{Ar} is proposed and discussed. Increased T_{Ar} is also shown to reduce the sp^3 hybridization content and defect densities in the buffer layer on n-type conductive substrates. Growth on semi-insulating substrates results in ordered buffer layer with significantly improved structural properties, for which T_{Ar} plays only a minor role. The free charge density and mobility parameters of monolayer graphene and quasi-freestanding monolayer graphene with different T_{Ar} and different environmental treatment conditions are determined by contactless terahertz optical Hall effect. An efficient annealing of donors on and near the SiC surface is suggested to take place for intrinsic monolayer graphene grown at 2000 °C, and which is found to be independent of T_{Ar}. Higher T_{Ar} leads to higher free charge carrier mobility parameters in both intrinsically n-type and ambient p-type doped monolayer graphene. T_{Ar} is also found to have a profound effect on the free hole parameters of quasi-freestanding monolayer graphene. These findings are discussed in view of interface and buffer layer properties in order to construct a comprehensive picture of high-temperature sublimation growth and provide guidance for growth parameters optimization depending on the targeted graphene application.

Keywords: epitaxial graphene on SiC; buffer layer; quasi-free-standing graphene; monolayer graphene; high-temperature sublimation; terahertz optical Hall effect; free charge carrier properties

1. Introduction

Epitaxial graphene on SiC substrates [1–4] holds promise for myriad of future electronic and sensing applications [5–9]. In particular, on the Si-face of SiC, the number of graphene layers can be well controlled and uniform monolayer graphene (MLG) can be obtained. Epitaxial graphene grown in ultra-high vacuum (UHV) on Si-face SiC consists of small domains with a typical size of 200–500 nm [10–15]. In such instances the surface roughens during the graphitization even when growth starts from an atomically-flat surface. If the graphitization is performed in argon (Ar) atmosphere, smoother surface and large-size MLG domains can be obtained [1,2,15]. However, small inclusions of bi-layer graphene (BLG) are typically present, most often formed on the step edges (due to the small miscut of nominally on-axis wafers) or in association with surface defects [15,16]. Hydrogen pre-treatment has widely been used to provide step-like surface morphology with atomically flat terraces and typical step height of 0.75 nm. Consequently, BLG always forms on the step edges of hydrogen etched SiC and giant step bunching is observed in the graphitization process [2,17]. The two layers in the BLG are AB-stacked, hence possessing a parabolic band structure in contrast to the linearly dispersing bands (Dirac cones) at the K points of the first Brillouin zone of MLG. As a result, BLG inclusions may degrade significantly the transport properties of graphene on the Si-face of SiC and limit its applications [18,19].

Several approaches dispensing with H etching have been explored to eliminate giant step bunching. For example, we have shown that high-temperature sublimation ($T >$ 1800 °C) in Ar atmosphere in closed graphite crucible delivers wafer-scale MLG with negligible BLG inclusions and without hydrogen pre-treatment [1,15,20–24]. Other open-reactor strategies involve pre-conditioning of the SiC wafer by annealing in Ar and/or use of polymer layer, which enables smooth and uniform BLG-free MLG [4,17,25].

Formation of MLG on the Si-face SiC is preceded by consecutive surface reconstructions as the wafer is heated up [26]. The surface undergoes reconstruction from the Si-enriched (3 × 3) phase to the C-enriched ($6\sqrt{3} \times 6\sqrt{3}$)-R30° phase. The latter phase is often called "buffer layer" or "zero-layer" graphene because it has the same honeycomb lattice structure as graphene. About 1/3 of the C atoms in this initial layer are covalently bound to the SiC surface and thus the buffer layer is devoid of the electronic properties of graphene [27]. Hydrogen intercalation may be employed to decouple the buffer layer from the substrate turning it into quasi-free-standing (QFS) MLG as the former covalent bonds are broken and the Si dangling bonds at a SiC surface are saturated with hydrogen [27,28].

In UHV conditions the surface reconstructions up to the ($6\sqrt{3} \times 6\sqrt{3}$)-R30° phase occur in the temperature range of 800–1200 °C [29]. Upon heating to a higher temperature, the buffer layer decouples from the SiC to form a graphene sheet and another buffer layer forms underneath. Tropm and Hannon [26] have shown that the temperature range within which the surface reconstructions occurs can be shifted up by as much as 200 °C in comparison to the case of an ultrahigh vacuum by increasing the Si background pressure to $\sim 8 \times 10^{-7}$ Torr using disilane. Ar atmosphere efficiently enhances the Si pressure at the substrate surface since Ar atoms act as a diffusion barrier that limits the Si desorption from the surface. As a result, in Ar atmosphere graphene starts to form at higher temperatures as compared to growth in UHV. It has been shown that in an open Ar atmosphere with a pressure of ~ 900 mbar graphene starts to form at temperatures above 1550 °C and the buffer layer forms between 1400 °C and 1550 °C [2,4].

Forming the buffer layer at higher temperature has been theoretically suggested to be the key to grow high-quality graphene [30]. Experimentally it has also been shown that forming a smooth buffer layer at a temperature of $T \simeq 1400$ °C prevents giant step bunching and consequently it is possible to obtain a smooth surface covered with uniform MLG [17] even on wafers with a large miscut angle of 0.37°[4]. Introducing Ar at different temperatures during the graphitization process may provide an alternative pathway to influence the phase transition temperature between different surface reconstructions, and hence enable the growth of smooth MLG without the need of special pre-treatment. However, this approach has not been explored despite the intense investigation of buffer layer properties and optimization [4,31–34].

In this work, we report a comprehensive study of the effect of introducing Ar at different temperatures on the buffer layer formation and its properties in high-temperature sublimation for both n-type doped and high-purity semi-insulating (SI) 4H–SiC. The free charge carrier density and mobility parameters of the corresponding MLG and QFS-MLG are determined for different environmental conditions and discussed. A combined analysis of free charge carrier and structural properties provides insights into the graphitization processes in an enclosed environment and basis to design growth strategies depending on graphene targeted application.

2. Experimental Details

Buffer and MLG samples were prepared on the Si-face (0001) of on-axis SI and *n*-type doped 4H–SiC substrates (Cree, Inc., Durham, NC, USA) by high-temperature sublimation in Ar atmosphere [35] using the sublimation growth facilities at Linköping University. The thickness and miscut angle of the SI and *n*-type doped wafers were 360 μm and 0.09°, and 340 μm and 0.05°, respectively. The substrates were chemical–mechanical polished (CMP) on the Si-face and optically polished on the C-face. Samples with different sizes of 10 mm × 7 mm, 10 mm × 10 mm or 15 mm × 10 mm were fabricated. The substrates were first cleaned with acetone and ethanol, followed by the standard RCA1 and RCA2 cleaning procedures. Prior to transfer into the growth chamber, the substrates were treated with a hydrofluoric acid solution to remove the native oxide on the surface.

A graphite crucible with a closed inner cavity has been designed with the Virtual Reactor software (http://www.str-soft.com/products/Virtual_Reactor/ (accessed on 1 February 2021)) to provide uniform (within ∼0.5 °C) temperature distribution over 2-inch diameter wafer. The inner cavity design was optimized to minimize the lateral temperature variation resulting in a relatively complex shape. A sketch of the crucible is shown in Figure 1. A special graphite holder is used to position the SiC substrate in the crucible cavity. The crucible was placed into thermally-isolating porous graphite insulation and loaded into the growth chamber. The chamber is pumped down to vacuum level of ∼10^{-6} mbar and the crucible was inductively heated. Initially, the temperature is ramped up in vacuum at a rate of ∼16 °C per min until the crucible temperature, measured with pyrometer on its surface, has reached 1300 °C. During this initial temperature ramp-up, Ar gas with pressure P_{Ar} = 850 mbar was introduced into the chamber when the crucible temperature, T_{Ar}, was between 640 °C and 1300 °C. At the moment Ar was introduced the typical vacuum level was ∼5×10^{-5} mbar and it took about 5 min for the Ar pressure to reach P_{Ar} = 850 mbar. During this time the temperature typically increased by about 100°. Above 1300 °C, the temperature ramp-up continues at an increased rate of ∼70 °C per min until the targeted growth temperature, T_{gr}, is reached. The temperature is then kept constant for 0 min or 5 min, which we refer to as growth time, t_{gr}. During this final temperature ramp-up, P_{Ar} slightly increased to P_{Ar} = 880 mbar. Once the growth is finished, the inductive heating is switched off and the sample cools down passively at a rate of ∼65 °C per min. The MLG and buffer layer samples were grown at T_{gr} = 2000 °C and T_{gr} = 1600 °C, respectively. The growth conditions for all samples are listed in Table 1.

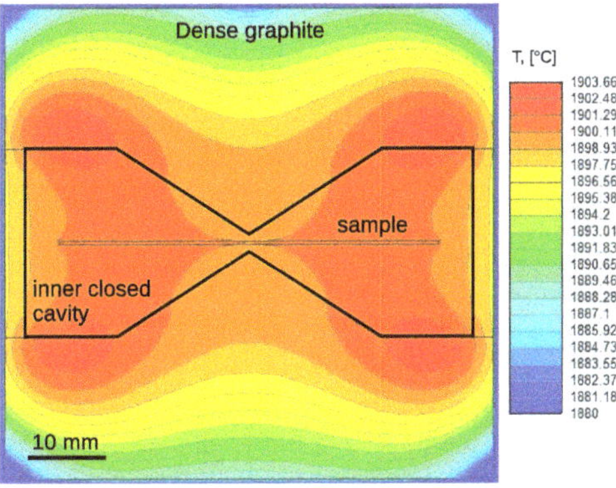

Figure 1. A schematic of the crucible with the distribution of the temperature overplotted. Note that the SiC substrate is placed within a tightly closed inner cavity and it is completely surrounded by graphite.

Table 1. Growth conditions of the samples studied in this work: T_{Ar}, T_{gr} and t_{gr} of buffer layer (BL), monolayer graphene (MLG) and quasi-free-standing (QFS)-MLG grown on n-type and semi-insulating (SI) substrates.

Sample	T_{Ar} [°C]	T_{gr} [°C]	t_{gr} [min]
	n-type 4H–SiC		
BL1	800	1600	0
BL2	900	1600	0
BL3	1150	1600	0
BL4	1300	1600	0
MLG0	800	2000	0
	SI 4H–SiC		
BL5	800	1600	0
BL6	1300	1800	0
MLG1	640	2000	0
MLG2	800	2000	0
MLG3	1300	2000	5
QFS-MLG1	640	1600	0
QFS-MLG2	800	1600	0
QFS-MLG3	1300	1800	0

Micro-reflectance and micro-Raman scattering spectroscopy (μ-RS) maps were measured using the set-up described in Ref. [36]. A diode-pumped semiconductor laser with a wavelength of 532 nm (photon energy E_L = 2.33 eV) was used for the excitation. The full-width at half-maximum (FWHM) of the focused laser spot is ∼0.4 μm using a 100× objective. Typically, 30 × 30 μm^2 reflectance maps with step sizes of 0.3 μm were measured at different locations of the sample. The typical size of the Raman maps was 10 × 10 μm^2. For each Raman spectrum, the micro-reflectance was also simultaneously measured. To obtain clean Raman spectra of MLG and buffer layers, a Raman spectrum of a bare 4H–SiC substrate was subtracted. Furthermore, all Raman spectra are normalized to the 4H–SiC substrate.

The surface morphology of the MLG and buffer layers was characterized by tapping mode atomic force microscopy (AFM) (Veeco Dimension 3100). Microprobe low-energy electron diffraction (μ-LEED), low energy electron microscopy (LEEM), X-ray photoelectron emission microscopy (XPEEM) and micro-focused X-ray photoelectron spectroscopy (micro-XPS) were used to investigate the structural properties and chemical composition of the buffer layer samples. The experiments were performed using the ELMITEC-LEEM III instrument at the I311 beamline of the MAX-Lab synchrotron radiation facility in Lund, Sweden.

Contactless terahertz (THz) cavity-enhanced (CE) optical Hall effect (OHE) measurements were performed for the determination of graphene-free charge carrier properties using the custom-built ellipsometry instrumentation at the THz Materials Analysis Center [37]. The OHE describes the magnetic field induced optical birefringence generated by free charge carriers under the influence of the Lorentz force, and can be measured by Mueller matrix ellipsometry [38]. The CE-OHE measurements were performed at room temperature by placing the sample on either of the two sides of a permanent neodymium magnet with a field strength of B = 0.548 T and an external cavity of ∼100 μm [39]. In-situ environmental control gas cell was employed to measure the samples in different gases and relative humidity (RH) [37,40]. Mueller Matrix data collected at magnetic fields B = +0.548 T and B = −0.548 T and their differences were simultaneously analyzed using a stratified optical model with parameterized model dielectric functions (MDFs) assigned to each layer, following the methodology described in Ref. [38]. The model consists of a perfect mirror (magnet), air gap, 4H–SiC substrate and an MLG or a QFS-MLG layer. The dielectric function of 4H–SiC was first determined from measurements of a bare substrate.

The substrate MDF parameters were then kept fixed during the analysis of the graphene samples. The MDF of graphene was described by Drude contribution in the presence of magnetic field [37,38]. The free charge carrier mobility μ and sheet density N_s of graphene were determined by non-linear least-squares fit of the calculated Mueller matrix data to the experimental data. The effective mass m^* was parametrized as $m^* = \sqrt{(h^2 N_s)/(4\pi v_F^2)}$ following Ref. [41], where $v_F = 1.02 \times 10^6 \text{ m s}^{-1}$ is the Fermi velocity and N_s is the carrier sheet density.

3. Results and Discussion

3.1. Buffer Layer Formation

Figure 2 shows μ-Raman spectra of buffer layers on *n*-type 4H–SiC, for which the Ar gas was introduced at $T_{Ar} = 800\ °C$, 900 °C, 1150 °C and 1300 °C, respectively (BL1-BL4, Table 1). The Raman spectra reveal features in the range of 1200–1700 cm^{-1}, typical for the buffer layer [31,33,42]. The band around 1330 cm^{-1} appears to be on par in terms of intensity with the band around 1580 cm^{-1} for all samples. It has been argued that the buffer layer Raman spectrum is not composed of discrete peaks but rather reflects the vibrational density of states [42]. The integrated intensity ratio of the D-band around 1330 cm^{-1} (D_{BL}) and the G-band 1580 cm^{-1} (G_{BL}) can be used to evaluate the content of sp^3 hybridization [31] or discuss correlations associated with buffer structure in general [33]. We will come back to this question when comparing buffer layers grown on *n*-type and SI 4H–SiC. However, what is important to the present discussion is the observation that the intensities of the two bands scale down with increasing T_{Ar} (see Figure 2). The analysis of the Raman scattering maps shows that the areas with lower reflectivity are associated with lower intensity of the D_{BL} and G_{BL} bands, which we attribute to lower buffer layer coverage. Furthermore, we estimate that the difference of the reflectance between regions that are barely covered with buffer and those with full coverage is ∼1%. Hence, reflectance mapping can also be employed to obtain information on the buffer layer uniformity on a large-scale.

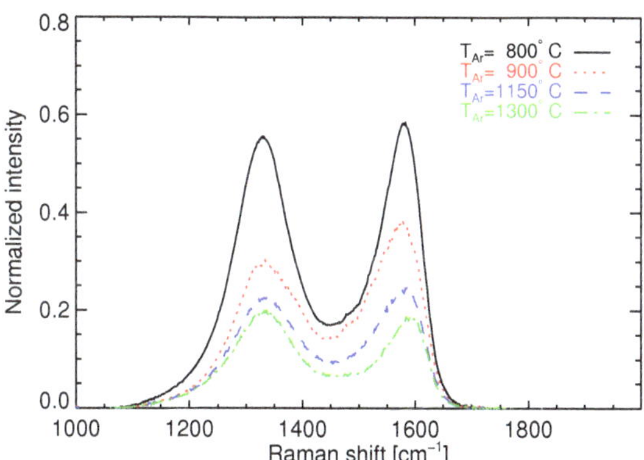

Figure 2. Normalized average μ-Raman scattering spectra obtained over 3 μm × 3 μm maps for the buffer layer samples with different T_{Ar}, indicated in the inset.

The μ-LEED patterns and the respective 30 μm × 30 μm reflectance maps of the buffer layer samples from Figure 2 are shown in Figure 3. The μ-LEED pattern of the sample with $T_{Ar} = 800\ °C$ (Figure 3a) displays well resolved $(6\sqrt{3} \times 6\sqrt{3})$-R30° surface reconstruction [11]. The uniform buffer layer coverage, for this sample, is corroborated by LEEM I(V) (not shown) and the reflectance map (Figure 3e), which reveals uniform intensity distribution. A clear buffer layer can also be inferred from the μ-LEED pattern of the buffer layer with $T_{Ar} = 900\ °C$ (Figure 3b), however, some charging on the surface is observed. The latter could be associated with oxidized SiC areas not covered by the buffer layer. For $T_{Ar} = 1150\ ^{\circ}C$ even stronger charging is observed in µ-LEED and patches of oxidized Si are identified by XPEEM (Figure 4). A mixture of the buffer layer and oxidized Si is inferred for this sample. Further confirmation of the suppressed buffer layer formation in the case of $T_{Ar} = 900\ ^{\circ}C$ and $T_{Ar} = 1150\ ^{\circ}C$ comes from the respective reflectance maps (Figure 3f,d), which show nonuniform intensity distribution with dark and bright areas. The size of the dark areas with suppressed buffer layer formation increases with increasing T_{Ar} up to 1150 °C. This sample also shows the highest RMS of 0.7 nm as compared to 0.35 nm and 0.5 nm for the buffer layers with $T_{Ar} = 800\ ^{\circ}C$ and $T_{Ar} = 900\ ^{\circ}C$, respectively. Note the resemblance between the XPEEM image (Figure 4) and the reflectance map (Figure 3g). We have previously reported a decrease in the relative reflectance of MLG with respect to the SiC substrate due to the presence of the oxide layer at the interface [43]. Finally, the sample with $T_{Ar} = 1300\ ^{\circ}C$ is severely charging and consists mostly of SiC substrate with the buffer layer just beginning to form, as revealed by µ-LEED (Figure 3d). In this case, the reflectance map (Figure 3h) appears quasi-uniform as the buffer layer nuclei are significantly smaller in comparison with the laser spot size.

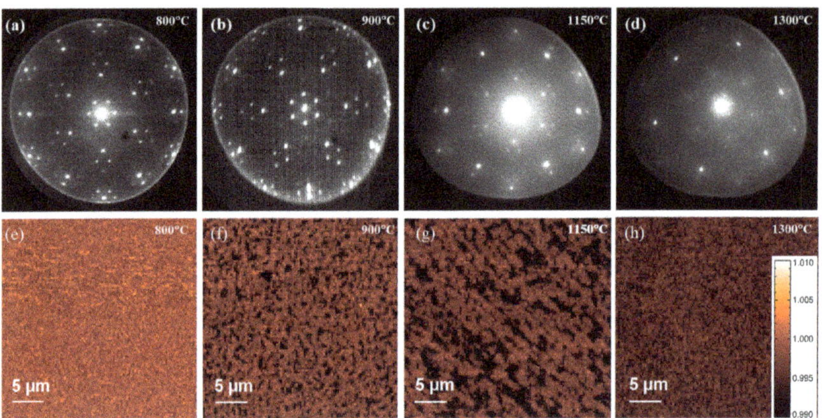

Figure 3. (**a–d**) Microprobe low-energy electron diffraction (µ-LEED) patterns taken at electron energy of 50 eV (**a,b**) and 40 eV (**c,d**) and (**e–h**) 30 µm × 30 µm normalized reflectance maps of buffer layer samples with $T_{Ar} = 800\ ^{\circ}C$ (**a,e**), $T_{Ar} = 900\ ^{\circ}C$ (**b,f**), $T_{Ar} = 1150\ ^{\circ}C$ (**c,g**) and $T_{Ar} = 1300\ ^{\circ}C$ (**d,h**). T_{Ar} are indicated in the up right corner of the respective images. The difference in reflectance between bare substrate and fully covered with buffer layer is ∼0.01.

Based on the Raman scattering spectroscopy, reflectance mapping as well as µ-LEED results, we can conclude that with the increasing temperature at which Ar is introduced, the formation of the buffer layer is suppressed and shifted to a higher temperature. The same trend is also consistently observed when the buffer layers are formed on SI 4H–SiC substrates. Our investigations further indicate that the SiC substrate areas not covered by the buffer layer are oxidized. There are three possible scenarios: (i) oxidation occurs after the buffer layer formation due to ambient exposure when the samples are removed from the reactor; (ii) oxidation occurs after the buffer layer formation during cooling down and (iii) oxidation occurs during the annealing process. Scenario (ii) and (iii) necessitate residual oxygen in the growth system. Oxidation of buffer and MLG samples as a result of residual oxygen has been previously observed for both conventional and high-temperature sublimation growth [44,45]. It has been suggested that since the graphitization process does not take place in ultra-high vacuum (oxygen-free) conditions, oxygen may be present as a result of oxygen-containing adsorbates on graphite parts and/or inner walls of the reactor. Different growth strategies to obtain high-quality MLG and/or buffer layer (e.g., for QFS-MLG applications) should be employed depending on whether scenario (i), (ii) or (iii) transpires.

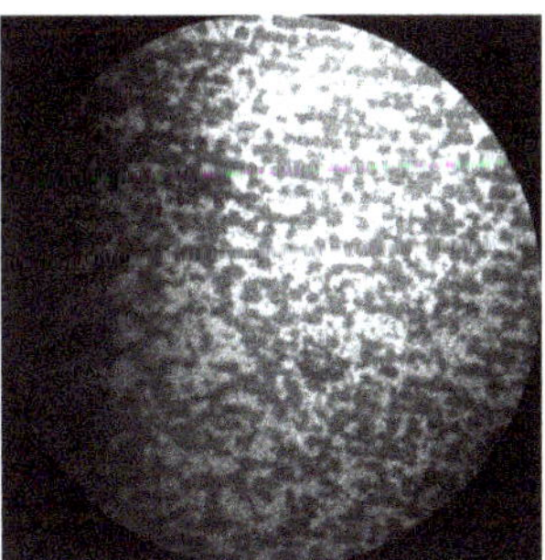

Figure 4. Si 2p oxide X-ray photoelectron emission microscopy (XPEEM) image taken at photon energy of 133 eV and electron energy of 26 eV with 40-µm field-of-view for the buffer layer sample with $T_{Ar} = 1150\,°C$. The bright areas correspond to higher content of SiO_x but even the dark areas of the image have some oxide component.

In order to elucidate which of the above scenarios takes place, we will discuss in the following the structural evolution of SiC during the sublimation process in Ar atmosphere. Both SiC restructuring and surface reconstruction are expected to be affected by the presence of Ar, which influences the gas pressure at the crystal-vapor interface and the mean free path length. Ar atmosphere effectively enhances the Si pressure since it leads to a reduced Si evaporation rate. The stability of steps on the SiC surface at a given temperature is also affected by Si pressure since the surface Si is in equilibrium with the gas phase Si as well as the bulk SiC. At higher Si pressures higher temperatures are needed to initiate Si decomposition from the terrace [30] and decomposition proceeds rather from the step resulting in smoother surface morphology as compared to ultrahigh vacuum [1,2]. Ar atmosphere also influences the mass transport of various species. Another consequence of the enhanced Si pressure in Ar is that Si depletion close to the SiC is slowed down and a higher temperature is needed to trigger and complete the buffer layer formation (consequently graphene formation). Indeed, it has been demonstrated that the phase transformation temperatures associated with different surface reconstructions on the Si-face SiC can be shifted by several hundred degrees Celsius by balancing the rate of Si evaporation with an external flux of Si [26]. In our experiments, when Ar is introduced at 800 °C the entire surface reconstruction process up to 1600 °C proceeds under enhanced Si pressure, which should shift the formation of the buffer layer to higher temperatures. In contrast, for $T_{Ar} = 1300\,°C$ the reconstruction occurs in vacuum up to this temperature and the formation of buffer layer should already take place [29]. We have previously shown that no etching by Ar occurs in the sublimation process in closed crucible [45] as confirmed here by step height distribution (See supplementary information Figure S1). Therefore, one would expect a better developed buffer layer for $T_{Ar} = 1300\,°C$ compared to $T_{Ar} = 800\,°C$. Surprisingly, we find the opposite trend from the Raman scattering spectroscopy, reflectance mapping and µ-LEED results. These findings are not compatible with scenarios (i) and (ii) in which oxidation of uncovered areas occurs after buffer layer formation. A potential explanation for the observed suppression of buffer layer formation at higher T_{Ar} is provided by scenario (iii) in which the observed oxidation occurs during the annealing process.

It has been shown that intermediate SiO_x on the Si-face of SiC is stable up to a temperature of 1200 °C and it is difficult to be fully eliminated even at 1400 °C [46]. Thus, if oxidation occurs during annealing and Ar is introduced at temperatures higher than 1200 °C the oxide layer will prevent the buffer layer formation. As the oxide layer starts to gradually be removed above 1200–1400 °C Ar effectively enhances the Si gas pressure and suppresses the phase transformation to $(6\sqrt{3} \times 6\sqrt{3})$-R30° surface reconstruction. As a results after heating up to 1600 °C, the sample with $T_{Ar} = 1300\,°C$ (BL4) shows only the initial stage of the buffer layer and is mostly uncovered SiC (Figure 3d). At T_{Ar} lower than 1200 °C (BL1, BL2, BL3), Ar reduces the mean free path of oxygen suppressing oxide formation and allowing complete (partial) buffer layer formation for $T_{Ar} = 800\ ^\circ C$ (900 $^\circ$C–1150 $^\circ$C.) We note that no charging or any indication of oxidation is observed in the buffer layer sample with $T_{Ar} = 800\ ^\circ C$, which may be understood in view of the reduced mean free path of oxygen at lower temperatures.

Scenario (iii) has several important implications for the growth strategies to obtain high-quality graphene by high-temperature sublimation. As the buffer layer becomes the first graphene layer upon annealing, forming the buffer layer and, consequently, graphene at higher temperatures should be favorable in terms of surface roughness and uniform restructuring as they affect positively free charge carrier mobility. At the same time, one can argue that if the buffer layer forms at lower temperatures it can be conditioned during the annealing process until the temperature of graphene formation is reached, reducing the density of defects such as vacancies or/and sp^3-defects. Another interesting question is to compare the properties of QFS-MLG obtained from buffer layers grown using different T_{Ar} and understand which mechanism has a decisive role. To address these questions we have investigated the free charge carrier properties of MLG and QFS-MLG samples for which the Ar was introduced at different T_{Ar} (Table 1). The MLG and QFS-MLG were grown on SI substrates in order to reliably measure the free charge carrier properties. Interestingly, a difference between the Raman scattering spectra grown at the same conditions on *n*-type and SI 4H–SiC is observed.

3.2. Comparison between Buffer Layers Grown on n-Type and SI 4H–SiC

A comparison of the Raman spectra of buffer layers on *n*-type and SI 4H–SiC obtained at $T_{Ar} = 800\ ^\circ C$ is presented in Figure 5a. The Raman spectrum of the buffer layer grown on n-type substrate displays D_{BL} (around 1330 cm^{-1}) and G_{BL} (around 1580 cm^{-1}) bands with similar intensities. The latter is slightly asymmetric due to a band at around 1530 cm^{-1} (see also Figure 2). Such Raman spectrum is typical for carbon-rich graphitic clusters bonded to SiC [27] and can be associated with a large degree of disorder [47]. On the other hand, the buffer layer grown at the same conditions but on SI substrates exhibits blue shift of the D_{BL} and the G_{BL} bands, and the band at around 1530 cm^{-1} becomes more pronounced. These are typical vibrational characteristics of a well-connected buffer layer domains [4]. Further information about disorder and the content of sp^3 hybridization can be obtained from the histograms of the G_{BL} band position (Figure 6a,c) and the ratios of the D_{BL} and G_{BL} bands areas, $A_{D_{BL}}/A_{G_{BL}}$, (Figure 6b,d). The G_{BL} band energy changes from 1583 cm^{-1} to 1606 cm^{-1} and the $A_{D_{BL}}/A_{G_{BL}}$ changes from 2.0 to 1.3 comparing the buffer layers grown on *n*-type and SI substrates, respectively. A similar trend is also found for the case of $T_{Ar} = 1300\ ^\circ C$ (Figure 6e,g). According to the amorphization trajectory presented for nano-crystalline graphite in Ref. [48], these changes can be associated with a significant reduction of the sp^3 hybridization content for the case of the SI 4H–SiC. The $A_{D_{BL}}/A_{G_{BL}}$ is further related to the degree of disorder introduced by the presence of sp^3 defects, which is proportional to the average distance between the defects [47]. Accordingly, the density of defects in the buffer layer grown on the SI substrate is 46% lower and the crystallite size is 35% larger. Again, very similar trend is found for the buffer layer with $T_{Ar} = 1300\ ^\circ C$ (Figure 6f,h). The observed differences between the two types of substrates could be understood considering the fact that electron concentration generally enhances thermal conductivity. Hence, temperature variations should occur slower for the SI substrates during the heating up, bringing the graphitization process closer to thermodynamic equilibrium and allowing the formation of a well-connected buffer layer with a lower density of defects. It is interesting to note that the vibrational features of the buffer layer formed underneath MLG, grown at $T_{gr} = 2000\ ^\circ C$ for 0 s, (Figure 5b) become even finer and bear closer resemblance with the buffer vibrational density of states [42]. Note that the spectral features are identical for the buffer layers on conductive and SI substrates. This further highlights the important roles of the carbon-rich environment and the high temperature for the formation of high-quality buffer layer.

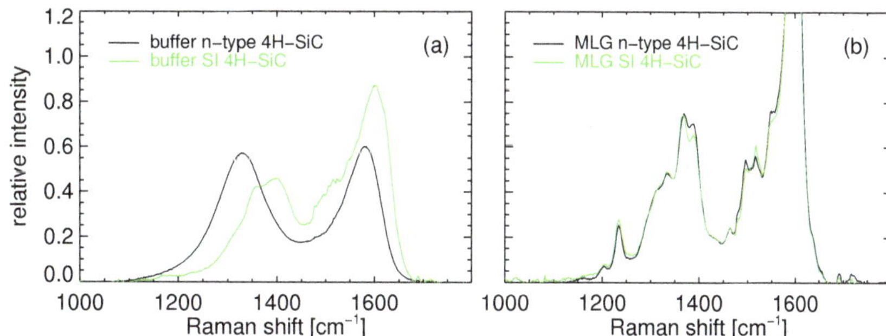

Figure 5. A comparison between the average μ-Raman scattering spectra for buffer layer samples with $T_{Ar} = 800\ °C$: (**a**) on n-type and SI 4H–SiC, and (**b**) the buffer layer features in fully-formed MLG at $T_{Gr} = 2000\ °C$ for 0 s on *n*-type and SI 4H–SiC.

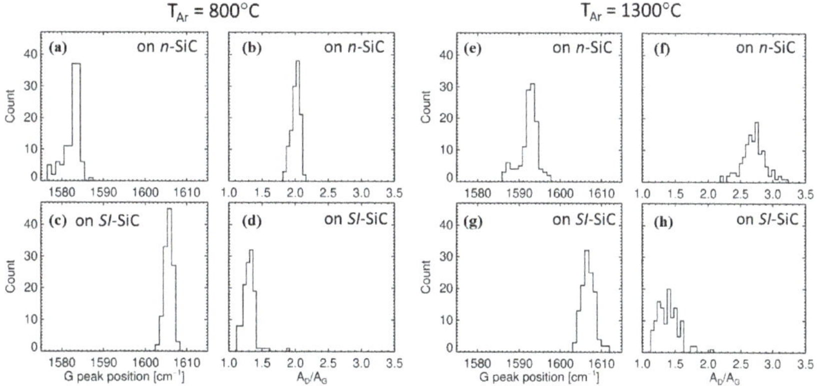

Figure 6. Histograms of the G_{BL} band position and the ratio of the D_{BL} and G_{BL} band areas, $A_{D_{BL}}/A_{G_{BL}}$, for the buffer layers grown with $T_{Ar} = 800\ °C$ (**a–d**) on *n*-type (**a,b**) and SI (**c,d**) 4H–SiC; and for the buffer layers grown with $T_{Ar} = 1300\ °C$ (**e–h**) on *n*-type (**e,f**) and SI (**g,h**) 4H–SiC. The histograms are obtained over Raman maps of 3 μm × 3 μm. Three Lorentzian lineshapes centered around G_{BL} of 1585–1600 cm^{-1}, D_{BL} of 1330–1530 cm^{-1} and a band centered at 1340 cm^{-1} were used for the fitting.

Comparing the buffer layers grown on *n*-type substrates and different T_{Ar}, a moderate blue-shift of the G-like band position for $T_{Ar} = 1300\ °C$ to 1593 cm^{-1} with respect to the sample with $T_{Ar} = 800\ °C$ (1583 cm^{-1}) can be seen (Figure 6a,e). This can be explained by a reduced sp^3 hybridization content as expected due to the higher temperature at which the reconstruction occurs. At the same time, the $A_{D_{BL}}/A_{G_{BL}}$ increases from 2.0 to 2.7 (Figure 6b,f), which could be related to a reduced crystallite size with 30%. This finding is in accordance with our μ-LEED results showing that the buffer layer with $T_{Ar} = 1300\ °C$ has just begun to form. We now turn our attention to the buffer layers grown with different T_{Ar} on SI 4H–SiC substrates. The same trend of suppressed reconstruction with increasing T_{Ar} is found. In fact, for the case of $T_{Ar} = 1300\ °C$ heating up to 1600 °C did not result into a buffer layer formation and heating up to 1800 °C was needed for a clear buffer layer Raman spectrum to be obtained. Interestingly, the buffer layers grown with $T_{Ar} = 800\ °C$ and $T_{Ar} = 1300\ °C$ exhibit very similar G_{BL} positions (Figure 6c,g) and $A_{D_{BL}}/A_{G_{BL}}$ ratios (Figure 6d,h), indicating similar sp^3 hybridization contents and densities of defects. A slightly broader distribution is observed for the case of $T_{Ar} = 1300\ °C$ for both n-type and SI 4H–SiC substrates, reflecting a slightly larger variation of the crystallite size.

Based on these results we can conclude that the temperature at which Ar is introduced has a determining role in the formation of the buffer layer in high-temperature sublimation in closed crucible independently of the 4H–SiC substrate conductivity. As a result of an interplay between oxidation and restructuring in Ar atmosphere, the formation of the buffer layer is shifted to higher temperatures for increased T_{Ar} of 1300 °C. Increasing T_{Ar} also leads to reduction of sp^3 hybridization contents and densities of defects on n-type 4H–SiC. However, T_{Ar} has a less pronounced effect for SI substrates, where ordered buffer layers form with similar structural properties.

3.3. Free Charge Carrier Properties of MLG and QFS-MLG

It is well-known that MLG on SiC is intrinsically n-type doped [49–51]. However, exposure to ambient can cause environmental doping of graphene via an acceptor redox reaction at the surface of the graphene involving various environmental gases (O_2, H_2O, and CO_2), which results in electron withdrawal [52]. Consequently, MLG can exhibit p-type conductivity depending on sample history [24,53]. We have previously shown that the THz OHE is an excellent tool to precisely determine free charge carrier density and mobility parameters of graphene and monitor their in-situ variation under the influence of different gases [24,37,40,54]. In order to determine the intrinsic properties of MLG and QFS-MLG, prior to the measurements they were annealed in vacuum (10^{-6} mbar) at 1000 °C and 500 °C, respectively. The annealing temperature was confirmed to not cause deintercalation or any changes in the QFS-MLG structural properties by LEEM, AFM, and µ-LEED. The samples were kept in dry N_2 during the measurements and storage. In addition, we have performed measurements after purging with dry N_2 for several days and air with RH of 45% for several hours. Both transient and static measurements were carried out. Finally, the samples were measured after being stored in ambient conditions for several months. We have selected for these investigations samples with the following T_{Ar}: (i) T_{Ar} = 800 °C, for which the surface reconstruction happens entirely in Ar atmosphere and that shows completed buffer layer after heating to T_{gr} = 1600 °C (0 s); (ii) T_{Ar} = 1300 °C, for which the surface reconstruction happens entirely in vacuum, and which needed heating to T_{gr} = 1800 °C for the buffer layer to form. Although no indications of surface oxidation were observed for the buffer layer sample with T_{Ar} = 800 °C, a nanoscale oxidation cannot be excluded. Furthermore, the graphitization process is shifted to higher temperatures in comparison to n-type substrate as pointed out above. We, therefore, included in our investigation MLG and QFS-MLG samples, for which the Ar was introduced at (iii) T_{Ar} = 640 °C. Growth temperature T_{gr} = 1600 °C was employed to produce the buffer layer sample in this case. The QFS-MLG samples were obtained by hydrogen intercalation of the respective buffer layers as described in Ref. [28]. The MLG samples were fabricated using our optimized conditions of T_{gr} = 2000 °C for 0 s growth time, which results in less than 1% BLG inclusions. The sample with T_{Ar} = 1300 °C required a longer growth time of 5 min for a homogeneous MLG to form leading to increased BLG inclusions of 8%.

Figure 7 shows the free charge carrier density (left panel) and mobility (right panel) of MLG (filled symbols) and QFS-MLG (open symbols) with different T_{Ar} for different environmental conditions. The mobility parameters were found to be slightly anisotropic in accordance with our recent study [24]. The anisotropy, which is caused by the substrate step edges, does not have any bearing on the results discussed in the current work. Consequently, for brevity we present here the averaged mobility between the parameters determined along and perpendicular to the step edge. The freshly annealed MLG samples show n-type conductivity, as expected, with values in the range of 3.9×10^{12} cm^{-2} to 6.6×10^{12} cm^{-2}. Due to the semi-insulating nature of the substrates, the MLG doping should be entirely governed by charge transfer due to surface donor states [55]. All three free electron density values are below the saturation density of n-type doping of MLG of 10^{13} cm^{-2} [55], indicating successful efficient annealing of donors on and near the SiC surface. The observed differences with T_{Ar}, albeit small, are significantly below the error bar of $0.3 \times$

10^{12} cm^{-2}. Since the MLG with $T_{Ar} = 1300\ ^\circ$C was obtained for a considerably longer time (5 min as compared to 0 s) it is tempting to speculate that the longer annealing may have a positive effect on reducing the interface dangling bonds effectively reducing the density of the surface state and leading to a lower free electron density. We have previously shown that purging with N$_2$ (or inert gases) effectively removes the ambient acceptor dopant, which may require up to several days of purging [37,40]. The free electron densities in the MLG samples with $T_{Ar} = 640\ ^\circ$C and $T_{Ar} = 800\ ^\circ$C after purging in dry N$_2$ for 9–10 days increased slightly to 5.1×10^{12} cm^{-2} and 7.0×10^{12} cm^{-2}, respectively, remaining below the the saturation density of n-type doping. The electron mobility parameters in these two cases slightly decreased in comparison to the freshly annealed samples, most likely as a result of the slightly increased charge density. The MLG with $T_{Ar} = 1300\ ^\circ$C shows the opposite behavior with slightly decreased charge density and slightly increased mobility parameter. Overall the purging with dry N$_2$ led to very small changes in the MLG electron density and mobility, which can be considered as the intrinsic free-electron parameters of MLG.

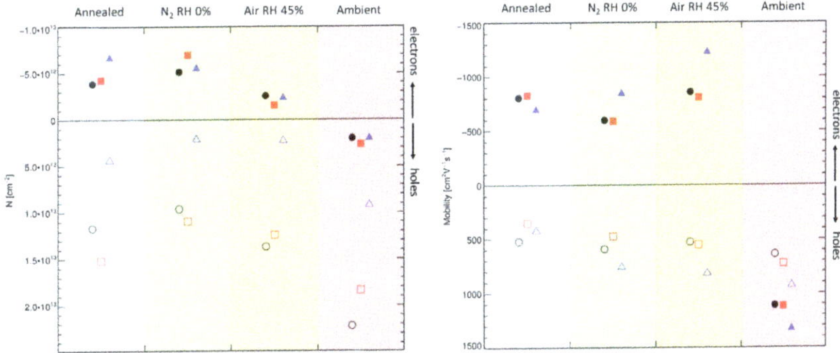

Figure 7. Free charge carrier density (left panel) and mobility (right panel) of MLG (filled symbols) and QFS-MLG (open symbols) with $T_{Ar} = 640\ ^\circ$C (black circles), $T_{Ar} = 800\ ^\circ$C (red squares) and $T_{Ar} = 1300\ ^\circ$C (blue triangles) for different environmental conditions: after annealing in vacuum (Annealed), after being purged with dry N$_2$ for several days (N$_2$ RH 0%), after being purged with moist air (RH 45%) for several hours (Air RH 45%), and after being exposed to the ambient for several months (Ambient).

As expected after purging with moist air (RH of 45%) the electron density in the MLG samples decreased due to the acceptor redox reaction at the graphene surface. The samples with different T_{Ar} show very similar electron density of $\sim 2 \times 10^{12}$ cm^{-2} after ~ 20 h of purging. We have measured the in-situ variations of free charge carrier properties and found that approximately 45 h purging in moist air are needed to flip the conductivity of MLG from *n*-type to *p*-type with free hole density of 1.4×10^{12} cm^{-2}. Long-term exposure in ambient conditions (several months) leads to only a very small increase of free hole density to $\sim 2 \times 10^{12}$ cm^{-2} indicating saturation of p-type ambient doping in MLG. Again, very similar free hole densities are found for the samples with different $T_{Ar} = 1300^\circ$. On the other hand, the free charge carrier mobility of the ambient doped MLG with $T_{Ar} = 1300\ ^\circ$C is more than 50% larger than the respective values of MLG with $T_{Ar} = 800\ ^\circ$C and $T_{Ar} = 640\ ^\circ$C. This is true for both the cases of free electrons and free holes (see Figure 7 right panel results for Air RH 45% and Ambient). This finding is very interesting considering that the samples with $T_{Ar} = 640\ ^\circ$C and $T_{Ar} = 800\ ^\circ$C have better MLG coverage of 99% and lower RMS $\simeq 0.4$ nm, compared with the $T_{Ar} = 1300\ ^\circ$C sample, which has 92% MLG coverage and RMS $\simeq 0.75$ nm. It was previously suggested that dominant scattering mechanisms at room temperature in graphene on SiC are the remote interface phonon scattering, as a result of coupling to the polar modes in the substrate, and scattering by impurities [56–58]. Since the MLG samples are grown at the same T_{gr} and have a similar history we do not anticipate a difference in impurity levels. It is thus plausible to suggest that in the MLG with T_{Ar} = 1300 °C the interface phonon scattering is reduced as a result of different interface properties. We recall that the buffer layers grown at different T_{Ar} on SI substrates exhibit very similar sp^3 contents and defect densities (Figure 6). Furthermore, the Raman scattering spectral features associated with the buffer layer in the respective MLG samples with different T_{Ar} are practically identical. Hence, the reduced interface phonon scattering is likely a result of a different interface between MLG and the buffer layer rather than between buffer layer and SiC substrate. This suggestion is further supported by the similar free charge carrier density in the ambient doped MLG with different T_{Ar} indicating similar surface state densities. To gain further insight into the origin of the different interface properties between MLG and the buffer layers we turn now our attention to the free charge carrier properties of the QFS-MLG samples.

In QFS-MLG the intercalated hydrogen saturates the Si dangling bonds passivating the interface donor states. Consequently, QFS-MLG exhibits p-type doping induced by the spontaneous polarization of the SiC substrate [28,59,60]. The resulting free hole density in QFS-MLG on SI 4H–SiC was reported to be 8.6×10^{12} cm^{-2} as determined by angular resolved photo-electron spectroscopy (ARPES) [60]. As expected our freshly annealed QFS-MLG samples show p-type conductivity (see Figure 7 left panel). We find very similar free hole densities in the QFS-MLG with T_{Ar} = 640 °C and T_{Ar} = 800 °C of 1.2×10^{13} cm^{-2} and 1.5×10^{13} cm^{-2}, respectively. These values are slightly higher than the free hole density expected from pure polarization doping [60]. It is possible that some residual ambient doping is present as the annealing temperature for the QFS-MLG samples was relatively low in order to prevent deintercalation. Purging in dry N$_2$ for several days lead to a small reduction of the free hole density in these two samples to $\sim 1.0 \times 10^{13}$ cm^{-2}, which is suggested to be the intrinsic value for our QFS-MLG resulting from polarization doping. We consider this to be a good agreement with the previously reported value of 8.6×10^{12} cm^{-2} [60] given the different experimental techniques used in the two works and the various fitting parameters employed to deduce the free hole concentration from ARPES. Both the freshly annealed and the dry N$_2$ purged QFS-MLG with T_{Ar} = 1300 °C show significantly lower free hole density of 4.4×10^{12} cm^{-2} and 2.1×10^{12} cm^{-2}, respectively. According to the polarization doping model, the negative pseudo-polarization charge, which is a constant parameter for the 4H–SiC, is balanced by the free holes in the QFS-MLG and the positive space charge in the substrate depletion layer [60]. Since the bulk doping in the SI substrate is the same for all three samples leading to a similar positive space charge in the substrate depletion layer, the observed lower free hole density in QFS-MLG with T_{Ar} = 1300 °C indicates the presence of donor surface states. As noted earlier, the buffer layers grown at different T_{Ar} on SI substrates exhibit very similar sp^3 contents and defect densities (Figure 6). We also confirmed by µ-Raman scattering spectroscopy mapping that no structural changes occur as a result of the intercalation process. Recall that in comparison to lower T_{Ar} the buffer layer with T_{Ar} = 1300 °C is incomplete. We speculate that this incomplete buffer layer formation may be the cause of the surface donor states, likely dangling bonds. Interestingly, purging with moist air (RH 45%) for $\sim$18 h leads to small increase of free hole density in QFS-MLG with T_{Ar} = 800 °C and T_{Ar} = 640 °C while for T_{Ar} = 1300 °C the hole density remains unchanged. This can be potentially explained by the above-mentioned scenario since the purge with moist air has different effects: for the polarization doped QFS-MLG it leads to chemical acceptor doping of graphene while for the sample with T_{Ar} = 1300 °C it leads to passivation of surface donor states. The two processes will naturally have different dynamics. This proposal is also consistent with the results for prolonged exposure to ambient. The free hole density in QFS-MLG with T_{Ar} = 1300 °C increases to 9.2×10^{12} cm^{-2} nearing the intrinsic polarization doping since most (all) surface donor states have been passivated. For T_{Ar} = 640 °C and T_{Ar} = 800 °C the free hole densities increase to 2.3×10^{13} cm^{-2} and 1.9×10^{13} cm^{-2}, respectively, as a result of chemical acceptor doping. In all cases, except for the freshly annealed samples, the largest hole mobility parameters are found for the QFS-MLG with T_{Ar} = 1300 °C . This is most likely related to the generally lower free hole density parameters. Note that the free charge mobility (and density) parameters represent average parameters obtained over the entire sample area of 10 mm × 10 mm.

4. Conclusions

We have critically reviewed the processes in high-temperature sublimation growth of graphene in Ar atmosphere using closed graphite crucible with emphasis on buffer layer formation and free charge carrier properties of MLG and QFS-MLG on 4H–SiC. We have explored the effect of introducing Ar at different temperatures, T_{Ar}. We have found that the buffer layer coverage decreases with increasing T_{Ar} with well-developed buffer layer for $T_{Ar} = 800\,°C$, while for $T_{Ar} = 1300\,°C$ the buffer layer is just beginning to form. The observed suppression of buffer layer formation at higher T_{Ar} is accompanied by surface oxidation of the uncovered regions of the SiC substrates. A scenario in which oxidation occurs during the annealing process is proposed to explain the peculiar shift of the buffer layer formation to higher temperatures. The latter leads to reduced sp^3 hybridization content and defect densities in the buffer layer when grown on n-type conductive substrates. Growth on SI substrates results in significantly improved structural properties of the buffer layers, which is attributed to a slower graphitization process closer to equilibrium due to the reduced thermal conductivity of the substrate. For SI substrate T_{Ar} plays a minor role for the sp^3 hybridization content and defect densities in the buffer layer. A comprehensive study of the free charge density and mobility parameters of MLG and QFS-MLG with $T_{Ar} = 640\,°C$, $T_{Ar} = 800\,°C$ and $T_{Ar} = 1300\,°C$ and four different environmental conditions: freshly annealed in vacuum, after purging with dry N_2 (RH 0%) for ~20 h, after purging with moist air (RH 45%) for ~18 h and after ambient exposure for several months, allows us to draw the following conclusions:

(i) successful efficient annealing of donors on and near the SiC surface can be inferred for MLG grown at 2000 °C independent of T_{Ar};

(ii) approximately 45 h purging with moist air (RH 45%) is needed to flip the conductivity of MLG from n-type to p-type and long term exposure to ambient leads to a saturation of the free hole density at ~2×10^{12} cm^{-2};

(iii) the highest mobility of MLG is determined for $T_{Ar} = 1300\,°C$ in both intrinsically n-type and ambient p-type doped situations. It is suggested that this is a result of reduced interface phonon scattering due to improved interface between MLG and the buffer layer rather than between the buffer layer and the SiC substrate;

(iv) a free hole density of ~1.0×10^{13} cm^{-2} is suggested to be the intrinsic value for our QFS-MLG resulting from polarization doping in good agreement with the previously reported value of 8.6×10^{12} cm^{-2} [60];

(v) T_{Ar} is found to have a profound effect on the free hole parameters of QFS-MLG. A significantly lower free hole density of ~2×10^{12} cm^{-2} is found in intrinsic QFS-MLG with $T_{Ar} = 1300\,°C$, which is attributed to additional surface donor states associated with incomplete buffer formation.

Our findings contribute to establishing a comprehensive picture of high-temperature sublimation growth and provide guidance for growth parameters optimization depending on the targeted application of QFS-MLG and MLG on SiC.

Supplementary Materials: The following are available online at https://www.mdpi.com/2076-3417/11/4/1891/s1, Figure S1: Representative AFM images and step height distributions of buffer layers grown on SI 4H–SiC with $T_{Ar} = 800\,°C$ and $T_{Ar} = 1300\,°C$.

Author Contributions: individual contributions of the authors are as follows: conceptualization, V.S. and V.D.; methodology, N.A. and P.K.; software, V.S.; validation, V.S.; formal analysis, V.S. and V.D.; investigation, V.S., N.A., A.A.Z., C.C., I.G.I. and C.B.; resources, R.Y. and V.D.; writing—original draft preparation, V.S. and V.D.; writing—review and editing, V.S., N.A., R.Y., C.C. and V.D.; visualization, V.S. and V.D.; supervision, V.D.; project administration, V.D.; funding acquisition, R.Y. and V.D. All authors have read and agreed to the published version of the manuscript.

Funding: The authors would like to acknowledge financial support from the Swedish Research Council (VR Contract 2016-00889), the Swedish foundation for strategic research (SSF) under Grants No. FFL12-0181 and No. RIF14-055, the Swedish Government Strategic Research Area in Materials Science on Functional Materials at Linköping University (Faculty Grant SFO Mat LiU No 2009 00971). RY is grateful for financial support by SSF via grant RMA 15-0024.

Acknowledgments: We thank Valdas Jokubavicius for his help with annealing the MLG and QFS-MLG samples in vacuum.

Conflicts of Interest: The authors declare no conflict of interest. The funders had no role in the design of the study; in the collection, analyses, or interpretation of data; in the writing of the manuscript, or in the decision to publish the results.

References

1. Virojanadara, C.; Syväjärvi, M.; Yakimova, R.; Johansson, L.I.; Zakharov, A.A.; Balasubramanian, T. Homogeneous large-area graphene layer growth on 6H-SiC(0001). *Phys. Rev. B* **2008**, *78*, 245403. [CrossRef]
2. Emtsev, K.V.; Bostwick, A.; Horn, K.; Jobst, J.; Kellogg, G.L.; Ley, L.; McChesney, J.L.; Ohta, T.; Reshanov, S.A.; Röhrl, J.; et al. Towards wafer-size graphene layers by atmospheric pressure graphitization of silicon carbide. *Nat. Mater.* **2009**, *8*, 203–207. [CrossRef]
3. de Heer, W.A.; Berger, C.; Ruan, M.; Sprinkle, M.; Li, X.; Hu, Y.; Zhang, B.; Hankinson, J.; Conrad, E. Large area and structured epitaxial graphene produced by confinement controlled sublimation of silicon carbide. *Proc. Natl. Acad. Sci. USA* **2011**, *108*, 16900–16905. [CrossRef] [PubMed]
4. Kruskopf, M.; Momeni Pakdehi, D.; Pierz, K.; Wundrack, S.; Stosch, R.; Dziomba, T.; Götz, M.; Baringhaus, J.; Aprojanz, J.; Tegenkamp, C.; et al. Comeback of epitaxial graphene for electronics: Large-area growth of bilayer-free graphene on SiC. *2D Mater.* **2016**, *3*, 041002, doi:10.1088/2053-1583/3/4/041002. [CrossRef]
5. Ang, P.K.; Chen, W.; Wee, A.T.S.; Loh, K.P. Solution-gated epitaxial graphene as pH sensor. *J. Am. Chem. Soc.* **2008**, *130*, 14392–14393. [CrossRef]
6. Lin, Y.M.; Dimitrakopoulos, C.; Jenkins, K.A.; Farmer, D.B.; Chiu, H.Y.; Grill, A.; Avouris, P. 100-GHz transistors from wafer-scale epitaxial graphene. *Science* **2010**, *327*, 662–662. [CrossRef]
7. Tzalenchuk, A.; Lara-Avila, S.; Kalaboukhov, A.; Paolillo, S.; Syväjärvi, M.; Yakimova, R.; Kazakova, O.; Janssen, T.J.B.M.; Falko, V.; Kubatkin, S. Towards a quantum resistance standard based on epitaxial graphene. *Nat. Nanotechnol.* **2010**, *5*, 186–189. [CrossRef]
8. Lin, Y.M.; Valdes-Garcia, A.; Han, S.J.; Farmer, D.B.; Meric, I.; Sun, Y.; Wu, Y.; Dimitrakopoulos, C.; Grill, A.; Avouris, P.; et al. Wafer-scale graphene integrated circuit. *Science* **2011**, *332*, 1294–1297. [CrossRef]
9. Rodner, M.; Bahonjic, J.; Mathisen, M.; Gunnarsson, R.; Ekeroth, S.; Helmersson, U.; Ivanov, I.G.; Yakimova, R.; Eriksson, J. Performance tuning of gas sensors based on epitaxial graphene on silicon carbide. *Mater. Des.* **2018**, *153*, 153–158.://doi.org/10.1016/j.matdes.2018.04.087. [CrossRef]
10. Seyller, T.; Emtsev, K.V.; Gao, K.; Speck, F.; Ley, L.; Tadich, A.; Broekman, L.; Riley, J.D.; Leckey, R.C.G.; Rader, O.; et al. Structural and electronic properties of graphite layers grown on SiC(0001). *Surf. Sci.* **2006**, *600*, 3906–3911. [CrossRef]
11. Riedl, C.; Starke, U.; Bernhardt, J.; Franke, M.; Heinz, K. Structural properties of the graphene-SiC(0001) interface as a key for the preparation of homogeneous large-terrace graphene surfaces. *Phys. Rev. B* **2007**, *76*, 245406. [CrossRef]
12. de Heer, W.A.; Berger, C.; Wu, X.; First, P.N.; Conrad, E.H.; Li, X.; Li, T.; Sprinkle, M.; Hass, J.; Sadowski, M.L.; et al. Epitaxial graphene. *Solid State Commun.* **2007**, *143*, 92–100. doi:10.1016/j.ssc.2007.04.023. [CrossRef]
13. Gu, G.; Nie, S.; Feenstra, R.M.; Devaty, R.P.; Choyke, W.J.; Chan, W.K.; Kane, M.G. Field effect in epitaxial graphene on a silicon carbide substrate. *Appl. Phys. Lett.* **2007**, *90*, 253507. [CrossRef]
14. Hibino, H.; Kageshima, H.; Maeda, F.; Nagase, M.; Kobayashi, Y.; Yamaguchi, H. Microscopic thickness determination of thin graphite films formed on SiC from quantized oscillation in reflectivity of low-energy electrons. *Phys. Rev. B* **2008**, *77*, 075413, doi:10.1103/PhysRevB.77.075413. [CrossRef]
15. Virojanadara, C.; Yakimova, R.; Zakharov, A.A.; Johansson, L.I. Large homogeneous mono-/bi-layer graphene on 6H-SiC(0001) and buffer layer elimination. *J. Phys. D Appl. Phys.* **2010**, *43*, 374010. [CrossRef]
16. Dimitrakopoulos, C.; Grill, A.; McArdle, T.J.; Liu, Z.; Wisnieff, R.; Antoniadis, D.A. Effect of SiC wafer miscut angle on the morphology and Hall mobility of epitaxially grown graphene. *Appl. Phys. Lett.* **2011**, *98*, 222105. [CrossRef]
17. Kruskopf, M.; Pierz, K.; Wundrack, S.; Stosch, R.; Dziomba, T.; Kalmbach, C.C.; Müller, A.; Baringhaus, J.; Tegenkamp, C.; Ahlers, F.J.; et al. Epitaxial graphene on SiC: Modification of structural and electron transport properties by substrate pretreatment. *J. Phys. Condens. Matter* **2015**, *27*, 185303. [CrossRef] [PubMed]
18. Ji, S.H.; Hannon, J.B.; Tromp, R.M.; Perebeinos, V.; Tersoff, J.; Ross, F.M. Atomic-scale transport in epitaxial graphene. *Nat. Mater.* **2012**, *11*, 114–119. [CrossRef] [PubMed]
19. Giannazzo, F.; Deretzis, I.; La Magna, A.; Roccaforte, F.; Yakimova, R. Electronic transport at monolayer-bilayer junctions in epitaxial graphene on SiC. *Phys. Rev. B* **2012**, *86*, 235422. [CrossRef]
20. Yakimova, R.; Yazdi, G.R.; Iakimov, T.; Eriksson, J.; Darakchieva, V. (Invited) Challenges of Graphene Growth on Silicon Carbide. *ECS Trans.* **2013**, *53*, 9–16. [CrossRef]

21. Darakchieva, V.; Boosalis, A.; Zakharov, A.A.; Hofmann, T.; Schubert, M.; Tiwald, T.E.; Iakimov, T.; Vasiliauskas, R.; Yakimova, R. Large-area microfocal spectroscopic ellipsometry mapping of thickness and electronic properties of epitaxial graphene on Si- and C-face of 3C-SiC(111). *Appl. Phys. Lett.* **2013**, *102*, 155411. [CrossRef]

22. Yakimova, R.; Iakimov, T.; Yazdi, G.; Bouhafs, C.; Eriksson, J.; Zakharov, A.; Boosalis, A.; Schubert, M.; Darakchieva, V. Morphological and electronic properties of epitaxial graphene on SiC. *Phys. B Condens. Matter* **2014**, *439*, 54–59. doi:10.1016/j.physb.2013.12.048. [CrossRef]

23. Yazdi, G.R.; Vasiliauskas, R.; Iakimov, T.; Zakharov, A.; Syväjärvi, M.; Yakimova, R. Growth of large area monolayer graphene on 3C-SiC and a comparison with other SiC polytypes. *Carbon* **2013**, *57*, 477. [CrossRef]

24. Armakavicius, N.; Kühne, P.; Eriksson, J.; Bouhafs, C.; Stanishev, V.; Ivanov, I.G.; Yakimova, R.; Zakharov, A.A.; Al-Temimy, A.; Coletti, C.; et al. Resolving mobility anisotropy in quasi-free-standing epitaxial graphene by terahertz optical Hall effect. *Carbon* **2021**, *172*, 248–259.://doi.org/10.1016/j.carbon.2020.09.035. [CrossRef]

25. Momeni Pakdehi, D.; Aprojanz, J.; Sinterhauf, A.; Pierz, K.; Kruskopf, M.; Willke, P.; Baringhaus, J.; Stöckmann, J.; Traeger, G.; Hohls, F.; et al. Minimum resistance anisotropy of epitaxial graphene on SiC. *ACS Appl. Mater. Interfaces* **2018**, *10*, 6039–6045. [CrossRef]

26. Tromp, R.M.; Hannon, J.B. Thermodynamics and Kinetics of Graphene Growth on SiC(0001). *Phys. Rev. Lett.* **2009**, *102*, 106104. [CrossRef]

27. Riedl, C.; Coletti, C.; Starke, U. Structural and electronic properties of epitaxial graphene on SiC(0001) : A review of growth, characterization, transfer doping and hydrogen intercalation. *J. Phys. D Appl. Phys.* **2010**, *43*, 374009. [CrossRef]

28. Riedl, C.; Coletti, C.; Iwasaki, T.; Zakharov, A.A.; Starke, U. Quasi-Free-Standing Epitaxial Graphene on SiC Obtained by Hydrogen Intercalation. *Phys. Rev. Lett.* **2009**, *103*, 246804. [CrossRef]

29. Forbeaux, I.; Themlin, J.M.; Debever, J.M. Heteroepitaxial graphite on $6H-SiC(0001)$: Interface formation through conduction-band electronic structure. *Phys. Rev. B* **1998**, *58*, 16396–16406. [CrossRef]

30. Kageshima, H.; Hibino, H.; Yamaguchi, H.; Nagase, M. Stability and reactivity of steps in the initial stage of graphene growth on the SiC(0001) surface. *Phys. Rev. B* **2013**, *88*, 235405. [CrossRef]

31. Strupinski, W.; Grodecki, K.; Caban, P.; Ciepielewski, P.; Jozwik-Biala, I.; Baranowski, J. Formation mechanism of graphene buffer layer on SiC(0001). *Carbon* **2015**, *81*, 63–72. [CrossRef]

32. Kruskopf, M.; Pierz, K.; Pakdehi, D.M.; Wundrack, S.; Stosch, R.; Bakin, A.; Schumacher, H.W. A morphology study on the epitaxial growth of graphene and its buffer layer. *Thin Solid Film.* **2018**, *659*, 7–15. [CrossRef]

33. Wang, T.; Huntzinger, J.R.; Bayle, M.; Roblin, C.; Decams, J.M.; Zahab, A.A.; Contreras, S.; Paillet, M.; Landois, P. Buffer layers inhomogeneity and coupling with epitaxial graphene unravelled by Raman scattering and graphene peeling. *Carbon* **2020**, *163*, 224–233. [CrossRef]

34. Momeni Pakdehi, D.; Pierz, K.; Wundrack, S.; Aprojanz, J.; Nguyen, T.T.N.; Dziomba, T.; Hohls, F.; Bakin, A.; Stosch, R.; Tegenkamp, C.; et al. Homogeneous Large-Area Quasi-Free-Standing Monolayer and Bilayer Graphene on SiC. *ACS Appl. Nano Mater.* **2019**, *2*, 844–852. doi:10.1021/acsanm.8b02093. [CrossRef]

35. Yakimova, R.; Iakimov, T.; Syväjärvi, M. Process for Growth of Graphene, PCT/SE2011/050328 (2011). U.S. Patent No. 9 150 417, 6 October 2015.

36. Ivanov, I.G.; Hassan, J.U.; Iakimov, T.; Zakharov, A.A.; Yakimova, R.; Janzén, E. Layer-number determination in graphene on SiC by reflectance mapping. *Carbon* **2014**, *77*, 492. [CrossRef]

37. Kühne, P.; Armakavicius, N.; Stanishev, V.; Herzinger, C.M.; Schubert, M.; Darakchieva, V. Advanced Terahertz Frequency-Domain Ellipsometry Instrumentation for In Situ and Ex Situ Applications. *IEEE Trans. Terahertz Sci. Technol.* **2018**, *8*, 257–270. [CrossRef]

38. Schubert, M.; Kühne, P.; Darakchieva, V.; Hofmann, T. Optical Hall effect-model description: Tutorial. *J. Opt. Soc. Am. A* **2016**, *33*, 1553. [CrossRef] [PubMed]

39. Knight, S.; Schöche, S.; Kühne, P.; Hofmann, T.; Darakchieva, V.; Schubert, M. Tunable cavity-enhanced terahertz frequency-domain optical Hall effect. *Rev. Sci. Instruments* **2020**, *91*, 083903, doi:10.1063/5.0010267. [CrossRef] [PubMed]

40. Knight, S.; Hofmann, T.; Bouhafs, C.; Armakavicius, N.; Kühne, P.; Stanishev, V.; Ivanov, I.G.; Yakimova, R.; Wimer, S.; Schubert, M.; et al. In-situ terahertz optical Hall effect measurements of ambient effects on free charge carrier properties of epitaxial graphene. *Sci. Rep.* **2017**, *7*, 1–8. [CrossRef]

41. Novoselov, K.S.; Geim, A.K.; Morozov, S.V.; Jiang, D.; Katsnelson, M.I.; Grigorieva, I.V.; Dubonos, S.V.; Firsov, A.A. Two-dimensional gas of massless Dirac fermions in graphene. *Nature* **2005**, *438*, 197–200. doi:10.1038/nature04233. [CrossRef] [PubMed]

42. Fromm, F.; Oliveira, M.H., Jr.; Molina-Sánchez, A.; Hundhausen, M.; Lopes, J.M.J.; Riechert, H.; Wirtz, L.; Seyller, T. Contribution of the buffer layer to the Raman spectrum of epitaxial graphene on SiC(0001). *New J. Phys.* **2013**, *15*, 043031, doi:10.1088/1367-2630/15/4/043031. [CrossRef]

43. Bouhafs, C.; Zakharov, A.A.; Ivanov, I.G.; Giannazzo, F.; Eriksson, J.; Stanishev, V.; Kühne, P.; Iakimov, T.; Hofmann, T.; Schubert, M.; et al. Multi-scale investigation of interface properties, stacking order and decoupling of few layer graphene on C-face 4H-SiC. *Carbon* **2017**, *116*, 722–732. [CrossRef]

44. Robinson, Z.R.; Jernigan, G.G.; Currie, M.; Hite, J.K.; Bussmann, K.M.; Nyakiti, L.O.; Garces, N.Y.; Nath, A.; Rao, M.V.; Wheeler, V.D.; et al. Challenges to graphene growth on SiC(000-1): Substrate effects, hydrogen etching and growth ambient. *Carbon* **2015**, *81*, 73–82. [CrossRef]

45. Jokubavicius, V.; Yazdi, G.R.; Ivanov, I.G.; Niu, Y.; Zakharov, A.; Iakimov, T.; Syväjärvi, M.; Yakimova, R. Surface engineering of SiC via sublimation etching. *Appl. Surf. Sci.* **2016**, *390*, 816–822. [CrossRef]

46. Schmeißer, D.; Batchelor, D.; Mikalo, R.; Hoffmann, P.; Lloyd-Spetz, A. Oxide growth on SiC(0001) surfaces. *Appl. Surf. Sci.* **2001**, *184*, 340–345. doi:10.1016/S0169-4332(01)00514-1. [CrossRef]

47. Martins Ferreira, E.H.; Moutinho, M.V.O.; Stavale, F.; Lucchese, M.M.; Capaz, R.B.; Achete, C.A.; Jorio, A. Evolution of the Raman spectra from single-, few-, and many-layer graphene with increasing disorder. *Phys. Rev. B* **2010**, *82*, 125429. [CrossRef]

48. Ferrari, A.C.; Robertson, J. Interpretation of Raman spectra of disordered and amorphous carbon. *Phys. Rev. B* **2000**, *61*, 14095–14107. [CrossRef]

49. Berger, C.; Song, Z.; Li, X.; Wu, X.; Brown, N.; Naud, C.; Mayou, D.; Li, T.; Hass, J.; Marchenkov, A.N.; et al. Electronic Confinement and Coherence in Patterned Epitaxial Graphene. *Science* **2006**, *312*, 1191–1196. [CrossRef]

50. Ohta, T.; Bostwick, A.; McChesney, J.L.; Seyller, T.; Horn, K.; Rotenberg, E. Interlayer Interaction and Electronic Screening in Multilayer Graphene Investigated with Angle-Resolved Photoemission Spectroscopy. *Phys. Rev. Lett.* **2007**, *98*, 206802. [CrossRef]

51. Emtsev, K.V.; Speck, F.; Seyller, T.; Ley, L.; Riley, J.D. Interaction, growth, and ordering of epitaxial graphene on SiC{0001} surfaces: A comparative photoelectron spectroscopy study. *Phys. Rev. B* **2008**, *77*, 155303. [CrossRef]

52. Sidorov, A.N.; Gaskill, K.; Buongiorno Nardelli, M.; Tedesco, J.L.; Myers-Ward, R.L.; Eddy, C.R.; Jayasekera, T.; Kim, K.W.; Jayasingha, R.; Sherehiy, A.; et al. Charge transfer equilibria in ambient-exposed epitaxial graphene on $(000 - 1)$ 6H-SiC. *J. Appl. Phys.* **2012**, *111*, 113706, doi:10.1063/1.4725413. [CrossRef]

53. Tedesco, J.L.; VanMil, B.L.; Myers-Ward, R.L.; McCrate, J.M.; Kitt, S.A.; Campbell, P.M.; Jernigan, G.G.; Culbertson, J.C.; Eddy, C.R.; Gaskill, D.K. Hall effect mobility of epitaxial graphene grown on silicon carbide. *Appl. Phys. Lett.* **2009**, *95*, 235406. [CrossRef]

54. Armakavicius, N.; Bouhafs, C.; Stanishev, V.; Kühne, P.; Yakimova, R.; Knight, S.; Hofmann, T.; Schubert, M.; Darakchieva, V. Cavity-enhanced optical Hall effect in epitaxial graphene detected at terahertz frequencies. *Appl. Surf. Sci.* **2017**, *421*, 357–360.://dx.doi.org/10.1016/j.apsusc.2016.10.023. [CrossRef]

55. Kopylov, S.; Tzalenchuk, A.; Kubatkin, S.; Fal'ko, V.I. Charge transfer between epitaxial graphene and silicon carbide. *Appl. Phys. Lett.* **2010**, *97*, 112109, doi:10.1063/1.3487782. [CrossRef]

56. Fratini, S.; Guinea, F. Substrate-limited electron dynamics in graphene. *Phys. Rev. B* **2008**, *77*, 195415. [CrossRef]

57. Tanabe, S.; Sekine, Y.; Kageshima, H.; Nagase, M.; Hibino, H. Carrier transport mechanism in graphene on SiC(0001). *Phys. Rev. B* **2011**, *84*, 115458. [CrossRef]

58. Lisesivdin, S.; Atmaca, G.; Arslan, E.; Çakmakyapan, S.; Kazar, Ö.; Bütün, S.; Ul-Hassan, J.; Janzén, E.; Özbay, E. Extraction and scattering analyses of 2D and bulk carriers in epitaxial graphene-on-SiC structure. *Phys. E Low-Dimens. Syst. Nanostructures* **2014**, *63*, 87–92. [CrossRef]

59. Ristein, J.; Mammadov, S.; Seyller, T. Origin of Doping in Quasi-Free-Standing Graphene on Silicon Carbide. *Phys. Rev. Lett.* **2012**, *108*, 246104, doi:10.1103/PhysRevLett.108.246104. [CrossRef]

60. Mammadov, S.; Ristein, J.; Koch, R.J.; Ostler, M.; Raidel, C.; Wanke, M.; Vasiliauskas, R.; Yakimova, R.; Seyller, T. Polarization doping of graphene on silicon carbide. *2D Mater.* **2014**, *1*, 035003. [CrossRef]

Article

Raman 2D Peak Line Shape in Epigraphene on SiC

Jan Kunc * and Martin Rejhon

Institute of Physics, Faculty of Mathematics and Physics, Charles University, Ke Karlovu 5,
CZ-121 16 Prague 2, Czech Republic
* Correspondence: kunc@karlov.mff.cuni.cz

Received: 29 February 2020; Accepted: 23 March 2020; Published: 30 March 2020

Abstract: We measured a 2D peak line shape of epitaxial graphene grown on SiC in high vacuum, argon and graphene prepared by hydrogen intercalation from the so called buffer layer on a silicon face of SiC. We fitted the 2D peaks by Lorentzian and Voigt line shapes. The detailed analysis revealed that the Voigt line shape describes the 2D peak line shape better. We have determined the contribution of the homogeneous and inhomogeneous broadening. The homogeneous broadening is attributed to the intrinsic lifetime. Although the inhomogeneous broadening can be attributed to the spatial variations of the charge density, strain and overgrown graphene ribbons on the sub-micrometer length scales, we found dominant contribution of the strain fluctuations. The quasi free-standing graphene grown by hydrogen intercalation is shown to have the narrowest linewidth due to both homogeneous and inhomogeneous broadening.

Keywords: epitaxial graphene; silicon carbide; Raman spectroscopy; 2D peak line shape; G peak; charge density; strain

---

1. Introduction

The Raman spectroscopy of graphene is a well-established technique [1–3] to determine number of graphene layers [4], strain [5,6], charge density [6–9], grain size [10–14], graphene functionalization [15], misorientation of graphene layers [16] or degree of hydrogen intercalation of epitaxial graphene on SiC [17,18]. The graphene's most prominent Raman spectral features are the D peak, G peak and 2D peak. The D peak reflects the amount of defects, or the graphene grain size. The G peak is related to the in-plane bond-stretching optical vibrations of sp^2 hybridized carbon atoms in the graphene lattice [19,20]. The 2D peak is recognized as a combination mode of lattice and electronic excitations. The predicted unique property of the 2D peak is its line shape. The 2D peak is predicted to have a Lorentzian line shape in the case of the single layer graphene (SLG) [21]. The 2D peak is predicted to have four components in the case of bilayer graphene. The single and four-component nature of the 2D peak was proved experimentally [22,23]. It is also known that the spectral position of the 2D peak is determined by the uniaxial [24–27] and biaxial [5,6,26] mechanical strain and charge density [7,8] of the graphene layer. The scaling of the 2D peak position with charge and strain was studied extensively. The 2D peak line shape is influenced by the strain uniformity [28] and strain fluctuations on the nanometer scale [29]. Also, different line shape was identified for graphene on SiC terraces and step edges [30].

However, beside the known spectral line shape and parameters determining the position of the 2D peak, there is little known about the combined effect of the homogeneous and inhomogeneous broadening. The knowledge of the mutual effects of the homogeneous and inhomogeneous broadening can provide deeper insight into the formation of the graphene layers and it can be used to further optimize graphene growth. Hence, we propose here to analyze in detail the spectral line shape of the 2D peak. Instead of describing the 2D peak line shape only by the Lorentzian broadening, we assume also the contribution by the inhomogeneous broadening. The homogeneous and inhomogeneous broadening have to be taken into account simultaneously. As the inhomogeneous broadening describes the random nature of the parameters determining the position of 2D peak, it is described by the normal (Gaussian) distribution. The mutual effect of the homogeneous and inhomogeneous broadening thus leads to the convolution of the Lorentzian and Gaussian broadening, also called the Voigt broadening. We show here that the Voigt profile describes the 2D peak line shape better than the Lorentzian profile. We show that the Voigt broadening, though it has one more fitting parameter, it does not show any signatures of over-parametrized model. We test the better fit quality by the F-test and the results are compared for three different samples and four different locations on each sample.

2. Materials and Methods

The epitaxial graphene was grown on a Si-face of 6H-SiC by thermal decomposition [31–34]. We grew three samples, each under different growth conditions. The sample grown in vacuum [31,32] at 10^{-5} mbar was heated to 1600 °C for 5 min. The sample grown in argon [33] at 1050 mbar was heated at 1650 °C for 5 min. The quasi free-standing monolayer graphene (QFMLG) [35,36] was grown in two steps. First, the graphene buffer layer was grown in 1050 mbar of argon at 1550 °C for 5 min. The second growth step was the hydrogen intercalation at 1120 °C for 5 min followed by a 2 hours long cooling to 600 °C. The growth temperature was adjusted with respect to the sublimation rate of silicon from the heated SiC wafer. The growth in high vacuum allows high silicon sublimation rates, hence, the temperature is reduced to 1600 °C in contrast to growth in 1050 mbar of argon, where the growth temperature has to be increased by 50 °C to 1650 °C to promote silicon sublimation comparable to the sublimation rate in high vacuum. The growth of QFMLG requires to grow so called buffer layer first. The buffer layer grows at even lower temperature than the single layer graphene. For this reason, the first growth step is performed at 1050 mbar of argon at 1550 °C for 5 min. The buffer layer is the graphene lattice, where about 30% carbons are sp^3 bonded to the underlying SiC substrate [37]. This sp^3 bonding was switched into sp^2 bonding by hydrogen intercalation [17,38]. More details on graphene growth and hydrogen intercalation can be found in our previous works [17,18,39].

Raman spectra were measured by WITec alpha300 (WITec, Ulm, Germany) micro-Raman confocal microscope. The Raman spectra were excited by 532 nm laser light. We used 25 mW laser power and the laser spot diameter was 1 μm in the focal plane. The spectra were acquired in two 30 s accumulations to achieve low level of noise in the tails of the 2D peak. The spatial Raman maps were measured on the area 4×4 μm^2 and we accumulated Raman spectra twice 5 s to optimize lateral resolution, noise level and the total measurement time. The Atomic Force Microscopy (AFM) and Lateral Force Microscopy (LFM) were measured by WITec alpha300 (WITec, Ulm, Germany) AFM. We measured AFM and LFM in the contact mode, and, the scanned area was 20×20 μm^2.

The 2D peak was fitted by Lorentzian and Voigt line shapes. The Lorentzian line shape was taken in the form of Equation (1)

$$G_L = \frac{\gamma}{\pi[(x - x_0)^2 + \gamma^2]},$$

(1)

where x_0 is the spectral position of the peak, γ determines the width of the 2D peak. The Equation (1) is scaled by a factor I_0 describing the intensity of the Lorentzian peak. Parameter γ is related to the more experimentally accessible Full Width at Half Maximum (FWHM) of the Lorentzian peak f_L by Equation (2).

$$f_L = 2\gamma$$

(2)

The Voigt line shape is given by the convolution of the Lorentzian Equation (1) and Gaussian Equation (3) line shape

$$G_G = \frac{1}{\sqrt{2\pi}\sigma} e^{-\frac{(x-x_0)^2}{2\sigma^2}}.$$

(3)

The Gaussian broadening σ relates to the FWHM f_G by Equation (4)

$$f_G = 2\sigma\sqrt{2\ln(2)}. \tag{4}$$

As the convolution of the Gaussian and Lorentzian function is numerically demanding, we used a common approximation of the Voigt line shape [40,41] given by Equation (5)

$$G_V = I_0 \left[\eta G_L + (1-\eta)G_G\right], \tag{5}$$

where η is a function of the total FWHM f, f_G and f_L

$$\eta = 1.36603\frac{f_L}{f} - 0.47719\left(\frac{f_L}{f}\right)^2 + 0.11116\left(\frac{f_L}{f}\right)^3. \tag{6}$$

The total Voigt broadening f is given by Equation (7)

$$f = (f_G^5 + 2.69269 f_G^4 f_L + 2.42843 f_G^3 f_L^2 + 4.47163 f_G^2 f_L^3 + 0.07842 f_G f_L^4 + f_L^5)^{1/5}. \tag{7}$$

3. Results

We show in Figure 1 the typical Raman spectra of graphene grown in high vacuum, argon and the intercalated buffer, so called QFMLG. All spectra show the typical graphene characteristics as the D peak, G peak and 2D peak. The Raman spectra also show typical characteristics of the single-layer graphene. These are the ratio of the integrated 2D peak to G peak intensity larger than 2, and, the single-component line shape of the 2D peak. The characteristic distinction between the hydrogen intercalated and non-intercalated samples can be also found by the absence/presence of the peaks labeled (1), (2) and (3) in inset of Figure 1. These peaks were attributed to the buffer layer [18,42]. The hydrogenated (QFMLG) and non-hydrogenated (SLG-vac, SLG-Ar) samples can be also distinguished by a low intensity background in the spectral range of the D peak. This background was assigned to the buffer layer, too [42]. We observe rather similar Raman spectra of graphene grown in argon and high vacuum. Their differences are discussed in the following detailed analysis.

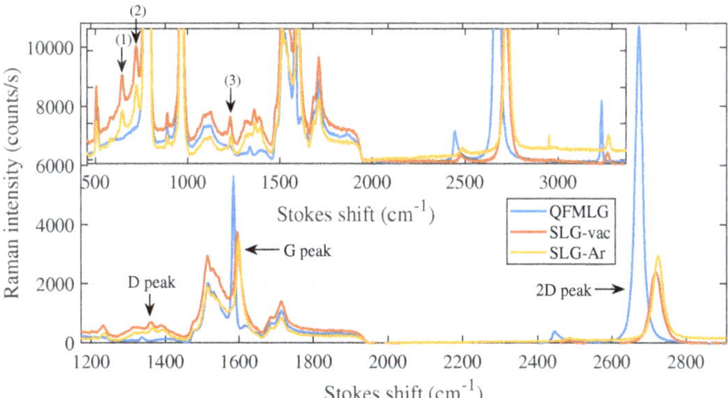

Figure 1. Raman spectra of quasi free-standing monolayer graphene (QFMLG), single layer graphene grown in vacuum (SLG-vac) and single layer graphene grown in argon (SLG-Ar). Graphene related D, G and 2D peaks are labeled by arrows. Inset shows details of Raman spectra including three buffer related Raman modes, labeled by arrows (1), (2) and (3).

The 2D peaks of the three samples are shown in Figure 2 by black circles. We obtained the spectra by two 30 s integration periods. The 30 s integration reduced the noise level and the two accumulations allowed us to remove spikes in recorded spectra. The low level of the noise is essential to fit the 2D peak line shape at low-energy and high-energy tails from the central peak position. Before we analyzed the 2D peak line shape, we subtracted a linear background. The background was determined using five experimental points at 2530 and at 2870 cm^{-1}. The 2D peaks are fitted by the Lorentzian (green curves) and Voigt (red curves) line shapes in Figure 2. The residuals are displayed in the top insets and the corresponding histograms of the residuals are shown in the bottom insets of Figure 2.

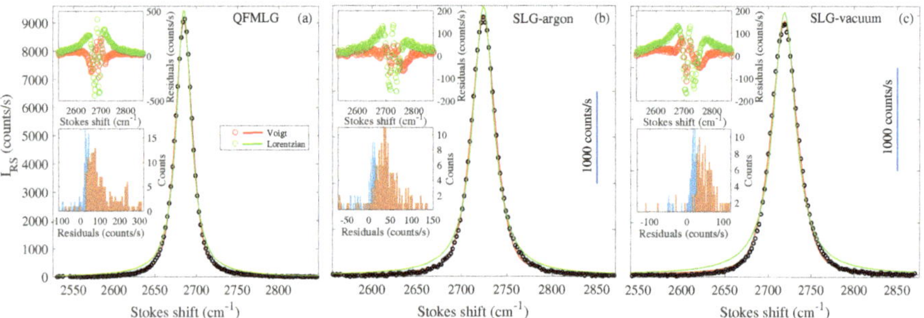

Figure 2. Line shape of the 2D peak in (**a**) QFMLG, (**b**) SLG-Ar and (**c**) SLG-vac. The black dots are the experimental data, green curve is a Lorentzian fit and red curve is the Voigt fit to the experimental data. The upper insets show the residuals for the (green point) Lorentzian and (red points) Voigt line shape. The bottom insets show the histograms of thee residuals.

The fit improvement by the Voigt line shape is tested by the F-test. The Degrees of Freedom (DOF) and Residual Sum of Squares of the Lorentzian (model 1) and Voigt (model 2) line shape are summarized in Table 1. The null hypothesis for the F-test is: the Lorentzian and Voigt line shapes are the same. The F-number in Table 1 results from the comparison of the two models. The p-number determines the cumulative probability that the expected F number is lower than the experimentally determined F-number.

Table 1. Parameters used to perform the F-test. Degrees of freedom (DOF), Residual Sum of Squares (RSS$_1$) of the Lorentzian model and RSS$_2$ of the Voigt model. F is the value of the F-statistics. Number *p* is the p-number to reject the null hypothesis.

Sample	DOF	RSS$_1$	RSS$_2$	F	p
QFMLG	163	3.6×10^6	4.9×10^5	1000	1
SLG-Ar	162	5.8×10^5	8.4×10^4	932	1
SLG-vac	163	8.0×10^5	1.5×10^5	675	1

The F-test shows that the Voigt line shape describes the 2D peak line shape better. As the Voigt line shape has four fitting parameters (one more with respect to the Lorentzian broadening), we need to verify that the Voigt line shape model is not overparametrized. We verify the overparametrization by determining the χ^2 statistics of the fitted models. The χ^2 statistics requires the experimental error σ_{exp}. We determine the experimental error from the high-energy tails of the 2D peak, as depicted in Figure 3. The high-energy tail is fitted by the second order polynomial to describe the trend of the experimental data. The residuals are considered as a random experimental error. The normality of these residuals is tested by the Kolmogorov-Smirnov test, as depicted by the cumulative distribution function in the inset of Figure 3. The normality is verified at the 95% confidence level. The experimental error is estimated to be $\sigma_{exp} \approx 10$ cm^{-1}. The centered and normalized χ^2 for the Lorentzian and Voigt model is 1000 and 260, respectively.

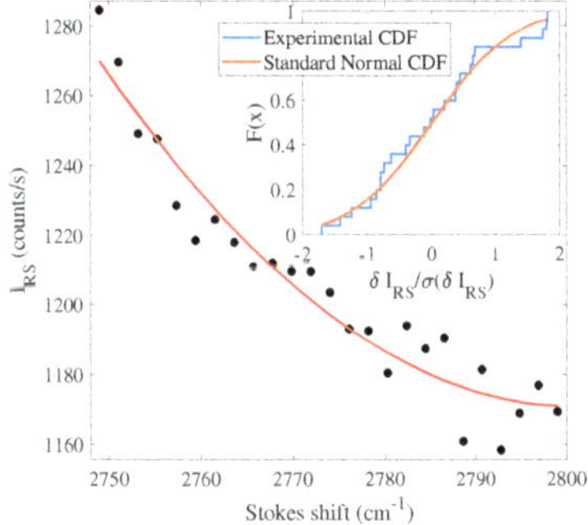

Figure 3. The high-energy tail of the 2D peak (black points) was fitted to the polynomial of the 2nd order (red curve) to determine the experimental error. The normality of the residuals is tested by the Kolmogorov-Smirnov test (inset).

We fitted the 2D peak spectra at four different positions on each sample. The statistics for the three samples is shown in Figure 4 and summarized in Table 2. The FWHM was obtained directly from the experimental data without using any fitting procedure. The parameters γ and σ are the fitting parameters of the Voigt line shape. We observe that the QFMLG shows the narrowest broadening. The QFMLG also shows the smallest contribution of the homogeneous and inhomogeneous broadening. It can be seen in Figure 4 that the homogeneous and inhomogeneous broadening are similar in QFMLG. However, the inhomogeneous broadening is larger by 2–3 cm^{-1} with respect to the homogeneous broadening in SLG-Ar and SLG-vac. The role of inhomogeneous broadening is thus larger in SLG-Ar and SLG-vac than in the QFMLG.

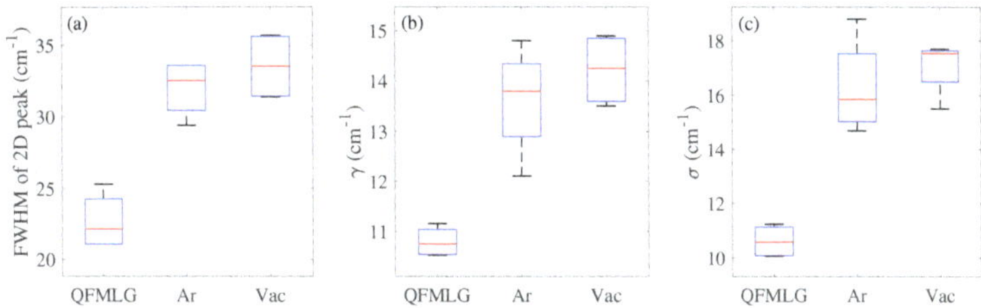

Figure 4. The (**a**) FWHM and (**b,c**) Voigt fitting parameters of (**b**) homogeneous (γ) and (**c**) inhomogeneous (σ) broadening are compared for the three samples (QFMLG, SLG-Ar, SLG-vac).

Table 2. Fitted parameters γ and σ of Voigt line shape and FWHMs for three samples (QFMLG, SLG-Ar, SLG-vac). The line shape was determined at four different positions on each sample.

Sample	Sample Position	FWHM (cm^{-1})	γ (cm^{-1})	σ (cm^{-1})
QFMLG	1	21.1	10.53 ± 0.12	10.13 ± 0.12
	2	21.1	10.56 ± 0.13	10.06 ± 0.13
	3	25.3	11.16 ± 0.16	11.24 ± 0.15
	4	23.2	10.94 ± 0.17	11.06 ± 0.16
Argon grown	1	33.6	14.8 ± 0.4	16.3 ± 0.3
	2	31.5	13.7 ± 0.3	15.4 ± 0.3
	3	29.4	13.9 ± 0.2	14.7 ± 0.2
	4	33.6	12.1 ± 0.6	17.6 ± 0.5
Vacuum grown	1	35.7	14.9 ± 0.5	17.5 ± 0.4
	2	31.5	13.5 ± 0.3	15.5 ± 0.2
	3	35.6	14.8 ± 0.4	17.7 ± 0.4
	4	31.4	13.7 ± 0.3	16.4 ± 0.2

We study sources of inhomogeneous broadening by measuring spatial Raman maps. We show the Raman maps of the 2D peak position and FWHM of 2D peak in Figure 5. We select three representative points (marked by black, red and blue circles in Figure 5a–f) and Raman spectra taken at these points are plotted in Figure 5g–i. The SLG-vac sample is the least homogeneous. The broad and blue-shifted 2D peak (red spectrum in Figure 5g) is a fingerprint of bilayer graphene. We observe similar small areas of broad and blue-shifted 2D peak in QFMLG, too (red spectrum in Figure 5i). Though SLG-Ar sample appears to be more homogeneous than SLG-vac and QFMLG, this could be due to the larger steps of SiC, or, due to the specific area chosen for the Raman map. To verify further the origin of these inhomogeneities, we measured also AFM and LFM, see Figure 6. The topography of SLG-vac shows circular-like SiC terraces of 1–4 μm in diameter. The SLG-Ar and QFMLG show regular SiC step bunching. The regular terraces are 1–4 μm broad in SLG-Ar, and, they are 6–7 μm broad in QFMLG.

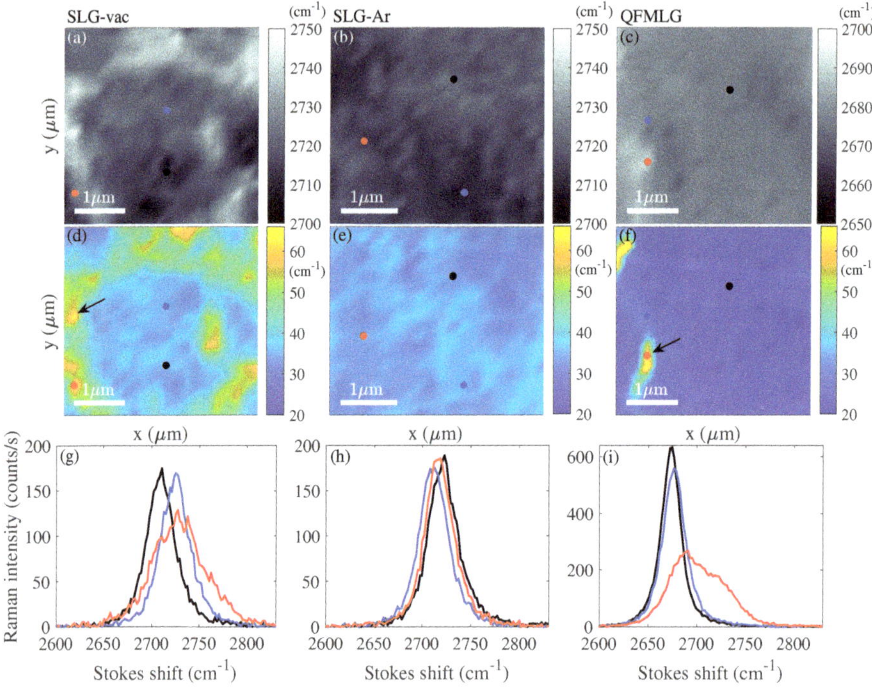

Figure 5. Maps of Raman scattering. The map of (**a**–**c**) 2D peak position and (**d**–**f**) 2D peak FWHM measured in (**a**,**d**,**g**) SLG-vac, (**b**,**e**,**h**) SLG-Ar and (**c**,**f**,**i**) QFMLG. We also show 2D peak line shape at three locations marked by black, red and blue circles. The spectra are plotted using corresponding black, red and blue curves in (**g**–**i**).

To identify these different areas in topography, we measured also LFM. The LFM can distinguish different materials if their friction with an AFM tip is different. The LFM images are shown in Figure 6d–f. We also plot friction force profiles in Figure 6g–i. We observe reduction of the friction force at the edges of homogeneous SiC areas in SLG-vac. Some edges show similar reduction of the friction force in SLG-Ar, too. In both cases, the friction force is reduced ≈ 2×. Such reduction of the friction force was shown between single and bilayer graphene [43]. The friction force is increased in the step edge areas in QFMLG. The increase of friction force was related to the buffer layer [43], however; we observe increase by only ≈ 2.5×. The expected increase is ≈ 10× for buffer layer. We note, that the step edge, or, sidewall area does not have the same structure as buffer, as shown in literature [44,45].

We interpret the graphene at the SiC step edges as the sidewall graphene ribbons [46,47]. We also observe in LFM that the friction displays large homogeneous areas, and, it also shows considerable amount of sub-micrometer sized inhomogeneities. These patches of different friction/material can contribute to the overall line shape of the 2D peak, too.

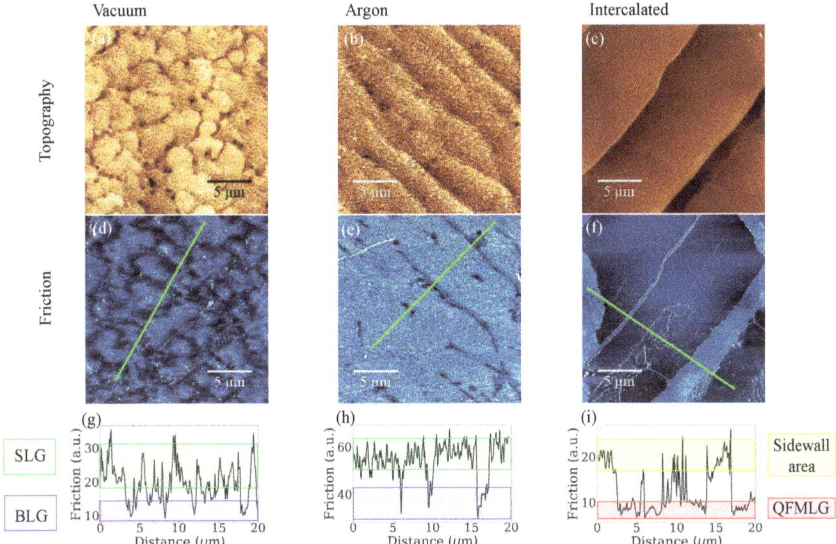

Figure 6. The topographies (20×20 µm^2) of investigated samples are depicted in graphs (**a–c**) and corresponding friction force maps are shown in graphs (**d–f**). The profiles (**g–i**) of friction force maps (green lines) demonstrate the SLG, BLG, QFMLG and sidewall area.

Another contribution to the 2D peak broadening is the charge density and strain variation on the sub-micrometer length scale. We analyzed the positions of the G and 2D peaks measured by Raman mapping. The correlation of the G and 2D peak positions are plotted in the inset of Figure 7. We describe the relation between the G and 2D peak position ω_G, ω_{2D} and mechanical strain and charge density by the following set of two equations.

$$\omega_G = \omega_{G0} + \alpha_G|n_{2D}| - 2\gamma_G\omega_{G0}\epsilon \tag{8}$$

$$\omega_{2D} = \omega_{2D0} - \alpha_{2D}n_{2D} - 2\gamma_{2D}\omega_{2D0}\epsilon, \tag{9}$$

where proportionality constants are $\alpha_G = 6.8$ cm$^{-1}/10^{13}$ cm^{-2} and $\alpha_{2D} = 2.7$ cm$^{-1}/10^{13}$ cm^{-2}. The effective Grüneisen parameters are $\gamma_G = 1.8$ and $\gamma_{2D} = 3.5$. The G and 2D peak positions in the charge neutral unstrained graphene are $\omega_{G0} = 1582$ cm^{-1}, and $\omega_{2D0} = 2680$ cm^{-1}. We chose the linear dependence of G and 2D peak position on strain in agreement with previous experimental works studying the G peak [7,26] and 2D peak [24–27]. We chose the functional dependence of the G peak position on the charge density as an approximation to the theoretically predicted dependence [7–9,48]. The expected charge density dependence of G peak position is smoothed at room temperature [48] in comparison to $T = 0$ K. The measured dependence can be clearly approximated by absolute-value function, when compared to theory [48] and experiment [5,7–9]. Our fitting parameters lead to 2.5× stronger charge density sensitivity of G peak than the sensitivity of 2D peak. The absolute values and relative strength of G to 2D peak charge sensitivity is in agreement with previous works [7,9]. Other fitting parameters lead to the sensitivity of G peak position to strain $\Delta\omega_G = -57$ cm$^{-1}/\%$, and, to the sensitivity of 2D peak position to strain $\Delta\omega_{2D} = -188 \text{ cm}^{-1}/\%$. These G and 2D peak sensitivities are in very good agreement with work of Mohiuddin [26], where authors found the G peak sensitivity to the biaxial strain $\Delta\omega_G = -63 \text{ cm}^{-1}/\%$ (page 4 in Ref. [26], 2nd column, 1st paragraph). The 2D peak sensitivity to the biaxial strain was found $\Delta\omega_G = -191 \text{ cm}^{-1}/\%$, see Ref. [26], page 5, 1st column, 4th paragraph. These parameters are also in good agreement with an experimental results of Schmidt [5] ($\Delta\omega_G = -63 \text{ cm}^{-1}/\%$, $\Delta\omega_{2D} = -149 \text{ cm}^{-1}/\%$), and, Density Functional Calculations (DFT) $\Delta\omega_G = -58 \text{ cm}^{-1}/\%$ and $\Delta\omega_{2D} = -144 \text{ cm}^{-1}/\%$, see Ref. [26], page 5, 2nd column.

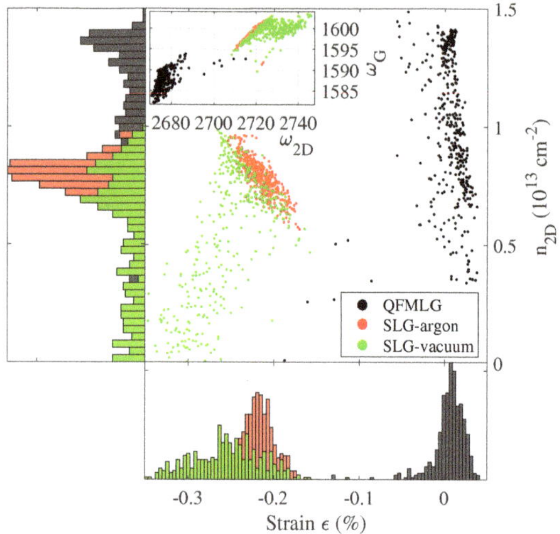

Figure 7. The variation of the charge density and strain as determined from the spectral position of the G and 2D peaks in the three samples (QFMLG, SLG-Ar, SLG-vac). The inset shows the correlation between the G and 2D peak positions. The left and bottom histograms depict the distribution of the charge density and strain, respectively.

We solve the set of Equations (8) and (9) numerically for each point of the Raman map and we obtain the correlation between the charge density and strain, as depicted in Figure 7. We observe that QFMLG shows the largest variation of the charge density. At the same time, the QFMLG sample shows the smallest inhomogeneous broadening, as we showed in spectral analysis of the 2D peak line shape fitted by the Voigt function. The anticorrelation between the observed charge density and inhomogeneous broadening in QFMLG led us to conclude that the main contribution to the sub-micrometer inhomogeneous broadening is the strain variation.

4. Discussion

The detailed analysis of 2D peak line shape shows that the Voigt line shape describes the 2D peak better than the Lorentzian line shape. Although the improvement is significant, the χ^2 statistics suggests that further corrections are needed. These deviations from the Voigt model are probably due to the larger areas of non-normally distributed strain, or, due to the patches of side wall graphene at the SiC step edges. Such areas cause side bands or local extrema in the normal distribution of the strain. We propose these deviations from the normal strain distribution can be reduced by defocusing the laser spot. However, defocusing will also reduce the Raman signal, and, it will also cause broader 2D peak. We assumed only biaxial strain contributing the total broadening of 2D peak. Though the uniaxial strain was neglected, it could contribute to the 2D peak broadening, too. The uniaxial strain splits the 2D peak [24–27]. If the 2D peak splitting is smaller or comparable to the spectral resolution (2 cm^{-1}), the 2D peak could be misinterpreted as a single-component peak. We assume that the uniaxial strain would effectively contribute to the 2D peak broadening only. Also, since the 2D peak splitting also leads to unequal intensities of the two split components, we expect contribution to the asymmetry of the 2D peak.

Another contribution to the inhomogeneity of epitaxial graphene on SiC is the stacking alignment between graphene and underlying SiC substrate [49]. These stacking domains can lead to strain inhomogeneity. However, since the size of these domains is well below the laser spot size (1 μm) we assume these stacking domains will mainly contribute to the Gaussian broadening, and, they could be studied by the Voigt line shape fitting.

We note it is necessary to fit the data of the 2D peak in the large enough spectral range. We used a $300\ \mathrm{cm}^{-1}$ broad spectral window. The spectral window broader than $300\ \mathrm{cm}^{-1}$ can not be used due to the presence of the combination Raman mode at $\approx 2450\ \mathrm{cm}^{-1}$. The narrower spectral range can lead to better fit [17,50], however, the fitting parameters might not be as reliable. As the Raman signal is low in the low and high energy tails, the signal might become weaker than the noise level. For the reason of high signal-to-noise ratio the Raman spectra have to be collected with a high numerical aperture objective and the data have to be integrated for at least 60 s.

Our analysis also shows a promise for growth improvements of the epitaxial graphene. The correlation between the width of the 2D peak and carrier mobility has been demonstrated [22]. Hence, we assume, if the strain inhomogeneity is reduced, the graphene carrier mobility could be improved, too.

Contrary to improving graphene, the effect of inhomogeneous broadening can also result in large deviations from the here presented line shapes. We have studied samples where the inhomogeneous broadening is comparable to the homogeneous broadening. However, if the inhomogeneous broadening dominates, the Gaussian component is expected to be prevalent in the 2D peak line shape.

5. Conclusions

We show the 2D peak line shape is given by a convolution of the inhomogeneous and homogeneous broadening, so called Voigt broadening, rather than just by a single Lorentzian line shape. We interpreted the inhomogeneous broadening to be mostly given by a sub-micrometer length scale strain variations. The hydrogen intercalated buffer layer is shown to have the smallest homogeneous and inhomogeneous 2D peak broadening.

Author Contributions: Individual contributions of the two authors are following: conceptualization, J.K.; methodology, J.K.; software, J.K. and M.R.; validation, J.K. and M.R.; sample preparation, J.K. and M.R.; formal analysis, J.K. and M.R.; investigation, J.K. and M.R.; resources, J.K.; data curation, J.K. and M.R.; writing—original draft preparation, J.K. and M.R.; visualization, J.K. and M.R.; supervision, J.K.; project administration, J.K.; funding acquisition, J.K. All authors have read and agreed to the published version of the manuscript.

Funding: This research was funded by Czech Science Foundation grant number 19-12052S. The MEYS project VaVpI CZ.1.05/4.1.00/16.0340 is also gratefully acknowledged.

Acknowledgments: We acknowledge helpful comments of J. Bok. These comments led us to the data analysis presented here.

Abbreviations

The following abbreviations are used in this manuscript:

SLG   single layer graphene
QFMLG   Quasi free-standing monolayer graphene
SLG-Ar   Single layer graphene grown in argon
SLG-vac   Single layer graphene grown in vacuum

AFM	Atomic Force Microscopy
LFM	Lateral Force Microscopy
DOF	Degrees of freedom
RSS	Residual Sum of Squares
FWHM	Full Width at Half Maximum
CDF	Cumulative distribution function

References

1. Tang, B.; Guoxin, H.; Gao, H. Raman Spectroscopic Characterization of Graphene. *Appl. Spectrosc. Rev.* **2010**, *45*, 369–407. [CrossRef]
2. Ferrari, A.C.; Basko, D.M. Raman spectroscopy as a versatile tool for studying the properties of graphene. *Nat. Nanotechnol.* **2013**, *8*, 235–246. [CrossRef] [PubMed]
3. Malard, L.M.; Pimenta, M.A.; Dresselhaus, G.; Dresselhaus, M.S. Raman spectroscopy in graphene. *Phys. Rep.* **2009**, *473*, 51–87. [CrossRef]
4. Park, J.S.; Reina, A.; Saito, R.; Kong, J.; Dresselhaus, G.; Dresselhaus, M.S. G ′ band Raman spectra of single, double and triple layer graphene. *Carbon* **2009**, *47*, 1303–1310. [CrossRef]
5. Schmidt, D.A.; Ohta, T.; Beechem, T.E. Strain and charge carrier coupling in epitaxial graphene. *Phys. Rev. B* **2011**, *84*, 235422. [CrossRef]
6. Nicolle, J.; Machon, D.; Poncharal, P.; Pierre-Louis, O.; San-Miguel, A. Pressure-Mediated Doping in Graphene. *Nano Lett.* **2011**, *11*, 3564–3568. [CrossRef]
7. Das, A.; Pisana, S.; Chakraborty, B.; Piscanec, S.; Saha, S.K.; Waghmare, U.V.; Novoselov, K.S.; Krishnamurthy, H.R.; Geim, A.K.; Ferrari, A.C.; et al. Monitoring dopants by Raman scattering in an electrochemically top-gated graphene transistor. *Nat. Nanotechnol.* **2008**, *3*, 210–215. [CrossRef]
8. Childres, I.; Jauregui, L.A.; Chen, Y.P. Raman spectra and electron-phonon coupling in disordered graphene with gate-tunable doping. *J. Appl. Phys.* **2014**, *116*, 233101. [CrossRef]
9. Yan, J.; Zhang, Y.; Kim, P.; Pinczuk, A. Electric field effect tuning of electron-phonon coupling in graphene. *Phys. Rev. Lett.* **2007**, *98*, 166802. [CrossRef]
10. Dresselhaus, M.S.; Jorio, A.; Souza Filho, A.G.; Saito, R. Defect characterization in graphene and carbon nanotubes using Raman spectroscopy. *Philos. Trans. R. Soc. A Math. Phys. Eng. Sci.* **2010**, *368*, 5355–5377. [CrossRef]
11. Liu, J.; Li, Q.; Zou, Y.; Qian, Q.; Jin, Y.; Li, G.; Jiang, K.; Fan, S. The Dependence of Graphene Raman D-band on Carrier Density. *Nano Lett.* **2013**, *13*, 6170–6175. [CrossRef] [PubMed]
12. Maultzsch, J.; Reich, S.; Thomsen, C. Double-resonant Raman scattering in graphite: Interference effects, selection rules, and phonon dispersion. *Phys. Rev. B* **2004**, *70*, 155403. [CrossRef]
13. Cancado, L.; Takai, K.; Enoki, T.; Endo, M.; Kim, Y.; Mizusaki, H.; Jorio, A.; Coelho, L.; Magalhaes-Paniago, R.; Pimenta, M. General equation for the determination of the crystallite size L-a of nanographite by Raman spectroscopy. *Appl. Phys. Lett.* **2006**, *88*, 163106. [CrossRef]
14. Cancado, L.G.; Jorio, A.; Martins Ferreira, E.H.; Stavale, F.; Achete, C.A.; Capaz, R.B.; Moutinho, M.V.O.; Lombardo, A.; Kulmala, T.S.; Ferrari, A.C. Quantifying Defects in Graphene via Raman Spectroscopy at Different Excitation Energies. *Nano Lett.* **2011**, *11*, 3190–3196. [CrossRef]
15. Niyogi, S.; Bekyarova, E.; Itkis, M.E.; Zhang, H.; Shepperd, K.; Hicks, J.; Sprinkle, M.; Berger, C.; Lau, C.N.; de Heer, W.A.; et al. Spectroscopy of Covalently Functionalized Graphene. *Nano Lett.* **2010**, *10*, 4061–4066. [CrossRef]
16. Kim, K.; Coh, S.; Tan, L.Z.; Regan, W.; Yuk, J.M.; Chatterjee, E.; Crommie, M.F.; Cohen, M.L.; Louie, S.G.; Zettl, A. Raman Spectroscopy Study of Rotated Double-Layer Graphene: Misorientation-Angle Dependence of Electronic Structure. *Phys. Rev. Lett.* **2012**, *108*, 246103. [CrossRef]
17. Kunc, J.; Rejhon, M.; Hlidek, P. Hydrogen intercalation of epitaxial graphene and buffer layer probed by mid-infrared absorption and Raman spectroscopy. *AIP Adv.* **2018**, *8*, 045015. [CrossRef]
18. Rejhon, M.; Kunc, J. ZO phonon of a buffer layer and Raman mapping of hydrogenated buffer on SiC(0001). *J. Raman Spectrosc.* **2019**, *50*, 465–473. [CrossRef]
19. Basko, D.M. Calculation of the Raman G peak intensity in monolayer graphene: role of Ward identities. *New J. Phys.* **2009**, *11*, 095011. [CrossRef]

20. Ferrari, A.C. Raman spectroscopy of graphene and graphite: Disorder, electron-phonon coupling, doping and nonadiabatic effects. *Solid State Commun.* **2007**, *143*, 47–57. [CrossRef]

21. Ferrari, A.C.; Meyer, J.C.; Scardaci, V.; Casiraghi, C.; Lazzeri, M.; Mauri, F.; Piscanec, S.; Jiang, D.; Novoselov, K.S.; Roth, S.; et al. Raman spectrum of graphene and graphene layers. *Phys. Rev. Lett.* **2006**, *97*, 187401. [CrossRef] [PubMed]

22. Robinson, J.A.; Wetherington, M.; Tedesco, J.L.; Campbell, P.M.; Weng, X.; Stitt, J.; Fanton, M.A.; Frantz, E.; Snyder, D.; VanMil, B.L.; et al. Correlating Raman Spectral Signatures with Carrier Mobility in Epitaxial Graphene: A Guide to Achieving High Mobility on the Wafer Scale. *Nano Lett.* **2009**, *9*, 2873–2876. [CrossRef] [PubMed]

23. Palmer, J.; Kunc, J.; Hu, Y.; Hankinson, J.; Guo, Z.; Berger, C.; de Heer, W.A. Controlled epitaxial graphene growth within removable amorphous carbon corrals. *Appl. Phys. Lett.* **2014**, *105*, 023106. [CrossRef]

24. Frank, O.; Mohr, M.; Maultzsch, J.; Thomsen, C.; Riaz, I.; Jalil, R.; Novoselov, K.S.; Tsoukleri, G.; Parthenios, J.; Papagelis, K.; et al. Raman 2D-Band Splitting in Graphene: Theory and Experiment. *ACS Nano* **2011**, *5*, 2231–2239. [CrossRef] [PubMed]

25. Huang, M.; Yan, H.; Heinz, T.F.; Hone, J. Probing Strain-Induced Electronic Structure Change in Graphene by Raman Spectroscopy. *Nano Lett.* **2010**, *10*, 4074–4079. [CrossRef]

26. Mohiuddin, T.M.G.; Lombardo, A.; Nair, R.R.; Bonetti, A.; Savini, G.; Jalil, R.; Bonini, N.; Basko, D.M.; Galiotis, C.; Marzari, N.; et al. Uniaxial strain in graphene by Raman spectroscopy: G peak splitting, Gruneisen parameters, and sample orientation. *Phys. Rev. B* **2009**, *79*, 205433. [CrossRef]

27. Yoon, D.; Son, Y.W.; Cheong, H. Strain-Dependent Splitting of the Double-Resonance Raman Scattering Band in Graphene. *Phys. Rev. Lett.* **2011**, *106*, 155502. [CrossRef]

28. Robinson, J.A.; Puls, C.P.; Staley, N.E.; Stitt, J.P.; Fanton, M.A.; Emtsev, K.V.; Seyller, T.; Liu, Y. Raman Topography and Strain Uniformity of Large-Area Epitaxial Graphene. *Nano Lett.* **2009**, *9*, 964–968. [CrossRef]

29. Neumann, C.; Reichardt, S.; Venezuela, P.; Droegeler, M.; Banszerus, L.; Schmitz, M.; Watanabe, K.; Taniguchi, T.; Mauri, F.; Beschoten, B.; et al. Raman spectroscopy as probe of nanometre-scale strain variations in graphene. *Nat. Commun.* **2015**, *6*, 8429. [CrossRef]

30. Grodecki, K.; Bozek, R.; Strupinski, W.; Wysmolek, A.; Stepniewski, R.; Baranowski, J.M. Micro-Raman spectroscopy of graphene grown on stepped 4H-SiC (0001) surface. *Appl. Phys. Lett.* **2012**, *100*, 261604. [CrossRef]

31. Hass, J.; de Heer, W.A.; Conrad, E.H. The growth and morphology of epitaxial multilayer graphene. *J. Phys. Condensed Matter* **2008**, *20*, 323202. [CrossRef]

32. De Heer, W.A.; Berger, C.; Ruan, M.; Sprinkle, M.; Li, X.; Hu, Y.; Zhang, B.; Hankinson, J.; Conrad, E. Large area and structured epitaxial graphene produced by confinement controlled sublimation of silicon carbide. *Proc. Natl. Acad. Sci. USA* **2011**, *108*, 16900–16905. [CrossRef] [PubMed]

33. Emtsev, K.V.; Bostwick, A.; Horn, K.; Jobst, J.; Kellogg, G.L.; Ley, L.; McChesney, J.L.; Ohta, T.; Reshanov, S.A.; Roehrl, J.; et al. Towards wafer-size graphene layers by atmospheric pressure graphitization of silicon carbide. *Nat. Mater.* **2009**, *8*, 203–207. [CrossRef] [PubMed]

34. Yazdi, G.R.; Iakimov, T.; Yakimova, R. Epitaxial Graphene on SiC: A Review of Growth and Characterization. *Crystals* **2016**, *6*, 53. [CrossRef]

35. Riedl, C.; Coletti, C.; Iwasaki, T.; Zakharov, A.A.; Starke, U. Quasi-Free-Standing Epitaxial Graphene on SiC Obtained by Hydrogen Intercalation. *Phys. Rev. Lett.* **2009**, *103*, 246804. [CrossRef] [PubMed]

36. Hassan, J.; Winters, M.; Ivanov, I.G.; Habibpour, O.; Zirath, H.; Rorsman, N.; Janzen, E. Quasi-free-standing monolayer and bilayer graphene growth on homoepitaxial on-axis 4H-SiC(0001) layers. *Carbon* **2015**, *82*, 12–23. [CrossRef]

37. Emtsev, K.V.; Speck, F.; Seyller, T.; Ley, L.; Riley, J.D. Interaction, growth, and ordering of epitaxial graphene on SiC{0001} surfaces: A comparative photoelectron spectroscopy study. *Phys. Rev. B* **2008**, *77*, 155303. [CrossRef]

38. Speck, F.; Jobst, J.; Fromm, F.; Ostler, M.; Waldmann, D.; Hundhausen, M.; Weber, H.B.; Seyller, T. The quasi-free-standing nature of graphene on H-saturated SiC(0001). *Appl. Phys. Lett.* **2011**, *99*, 122106. [CrossRef]

39. Kunc, J.; Rejhon, M.; Belas, E.; Dedic, V.; Moravec, P.; Franc, J. Effect of Residual Gas Composition on Epitaxial Growth of Graphene on SiC. *Phys. Rev. Appl.* **2017**, *8*, 044011. [CrossRef]

40. Ida, T.; Ando, M.; Toraya, H. Extended pseudo-Voigt function for approximating the Voigt profile. *J. Appl. Crystallogr.* **2000**, *33*, 1311–1316. [CrossRef]

41. Thompson, P.; Cox, D.; Hastings, J. Rietveld Refinement of Debye-Scherrer Synchrotron X-ray Data from Al_2O_3. *J. Appl. Crystallogr.* **1987**, *20*, 79–83. [CrossRef]

42. Fromm, F.; Oliveira, M.H., Jr.; Molina-Sanchez, A.; Hundhausen, M.; Lopes, J.M.J.; Riechert, H.; Wirtz, L.; Seyller, T. Contribution of the buffer layer to the Raman spectrum of epitaxial graphene on SiC(0001). *New J. Phys.* **2013**, *15*, 043031. [CrossRef]

43. Filleter, T.; McChesney, J.L.; Bostwick, A.; Rotenberg, E.; Emtsev, K.V.; Seyller, T.; Horn, K.; Bennewitz, R. Friction and Dissipation in Epitaxial Graphene Films. *Phys. Rev. Lett.* **2009**, *102*, 086102. [CrossRef] [PubMed]

44. Norimatsu, W.; Kusunoki, M. Formation process of graphene on SiC (0001). *Physica E* **2010**, *42*, 691–694. [CrossRef]

45. Berger, C.; Conrad, E.H.; de Heer, W.A. Epigraphene: Epitaxial graphene on silicon carbide. *arXiv* **2017**, arXiv:1704.00374.

46. Sprinkle, M.; Ruan, M.; Hu, Y.; Hankinson, J.; Rubio-Roy, M.; Zhang, B.; Wu, X.; Berger, C.; de Heer, W.A. Scalable templated growth of graphene nanoribbons on SiC. *Nat. Nanotechnol.* **2010**, *5*, 727–731. [CrossRef]

47. Baringhaus, J.; Aprojanz, J.; Wiegand, J.; Laube, D.; Halbauer, M.; Huebner, J.; Oestreich, M.; Tegenkamp, C. Growth and characterization of sidewall graphene nanoribbons. *Appl. Phys. Lett.* **2015**, *106*, 043109. [CrossRef]

48. Lazzeri, M.; Mauri, F. Nonadiabatic Kohn anomaly in a doped graphene monolayer. *Appl. Phys. Lett.* **2006**, *97*, 266407. [CrossRef]

49. De Jong, T.A.; Krasovskii, E.E.; Ott, C.; Tromp, R.M.; van der Molen, S.J.; Jobst, J. Intrinsic stacking domains in graphene on silicon carbide: A pathway for intercalation. *Phys. Rev. Mater.* **2018**, *2*, 104005. [CrossRef]

50. Pearce, R.; Tan, X.; Wang, R.; Patel, T.; Gallop, J.; Pollard, A.; Yakimova, R.; Hao, L. Investigations of the effect of SiC growth face on graphene thickness uniformity and electronic properties. *Surf. Topogr. Metrol. Properties* **2015**, *3*, 015001. [CrossRef]

Article

Structural Modifications in Epitaxial Graphene on SiC Following 10 keV Nitrogen Ion Implantation

Priya Darshni Kaushik [1,*], Gholam Reza Yazdi [1], Garimella Bhaskara Venkata Subba Lakshmi [2], Grzegorz Greczynski [1], Rositsa Yakimova [1] and Mikael Syväjärvi [1]

[1] Department of Physics, Chemistry and Biology, Linköping University, SE-58183 Linköping, Sweden; gholamreza.yazdi@liu.se (G.R.Y.); grzegorz.greczynski@liu.se (G.G.); rositsa.yakimova@liu.se (R.Y.); mikael.syvajarvi@liu.se (M.S.)

[2] Special Centre for Nanoscience, Jawaharlal Nehru University, New Delhi 110067, India; lakshmigbvs@gmail.com

* Correspondence: priya.kaushik@liu.se

Received: 29 April 2020; Accepted: 3 June 2020; Published: 10 June 2020

Abstract: Modification of epitaxial graphene on silicon carbide (EG/SiC) was explored by ion implantation using 10 keV nitrogen ions. Fragments of monolayer graphene along with nanostructures were observed following nitrogen ion implantation. At the initial fluence, sp^3 defects appeared in EG; higher fluences resulted in vacancy defects as well as in an increased defect density. The increased fluence created a decrease in the intensity of the prominent peak of SiC as well as of the overall relative Raman intensity. The X-ray photoelectron spectroscopy (XPS) showed a reduction of the peak intensity of graphitic carbon and silicon carbide as a result of ion implantation. The dopant concentration and level of defects could be controlled both in EG and SiC by the fluence. This provided an opportunity to explore EG/SiC as a platform using ion implantation to control defects, and to be applied for fabricating sensitive sensors and nanoelectronics devices with high performance.

Keywords: ion implantation; Raman; AFM; XPS; graphene

1. Introduction

Graphene is well known as a two-dimensional material with several remarkable properties like high electrical conductivity, high flexibility, low electric noise and high mechanical strength [1]. The lack of bandgap limits its application potential. Graphene based digital devices cannot be "switched-off" due to a constant flow of electrons. Switching is a necessity in electronic devices such as switches and transistors. Efforts are made to control the bandgap in growth and post processing, but it is a big challenge.

Modification of surfaces functionalizes graphene and has demonstrated a band gap [2], which makes possible its usage in digital electronics, nanoelectronics, spintronics, magnetronics and much more. The structural changes also alter its optical, electrical, mechanical and magnetic properties, thus offers its applicability for biosensors, gas sensors, optoelectronic devices, solar cells, detectors, batteries, pressure sensors, etc. The structure of graphene has been altered using various methods such as chemical doping, covalent functionalization and electron/ion beam irradiation [3–5].

The ion beam technology is a high precision and high reproducibility process. Structural modifications introduced using ion irradiation are much more effective due to the reason that there is an efficient transfer of energy from energetic incident ions to the atoms of graphene. Impurity doping can be done by ion irradiation (both by insertion of atoms from primary ion beam or via backscattered atoms from the substrate) and various atoms, molecules or ions can be explored. This range of options is not offered by any other technology. Both dopant and defects introduction in the material at the same time up to a desired depth can be also tailored.

The low energy ion implantation has been shown to modify the different types of graphene, such as produced by mechanical exfoliation, chemical vapor deposition (CVD) and sublimation epitaxy (epitaxial graphene) process. Experimental and theoretical studies showed an array of nanodots and nanowires on graphene on SiO_2 using 30 keV Ga ion irradiation [6]. The structural modification in mechanically exfoliated multilayer graphene on SiO_2 substrate caused an increase in defects with the increase in dwell time of ion irradiation, along with an increase in the compressive strain using focused ion beam (FIB) with Ga ions [7]. It was also shown that multilayered graphene is p-type doped due to defects thereby resulting in increasing the work function [7]. Modified single layer CVD graphene showed metal–insulator transition at larger doses of Ga ion irradiation using FIB [8]. There was a change of work function of graphene with an increase in defect density with increasing Ga irradiation dose [8]. Argon ion irradiation in the energy range of 0–200 eV with time of 0–10 s was applied on few layers of graphene. The study optimized beam energy and irradiation time to remove surface contaminants and to flatten the surface damage [9]. Nanopores were formed (size of two nanometers) following the treatment of exfoliated graphene on SiO_2 with hydrogen plasma etching [10]. Single layers of graphene nanopatterning were obtained by using helium ions and nanostructures and nanoribbons appeared at a dose of 2.0×10^{16} ions/cm^2 [11]. Nanopores were also obtained in exfoliated graphene using 35 keV Ga ion irradiation [12].

Different types of ions create different levels of defects. Xu et al. used B, N and F ion irradiation on CVD grown monolayer graphene at 35 eV and 20 eV energy between 1×10^{14} ions/cm^2 to 6×10^{15} ions/cm^2 fluence [13]. Willke and co-workers observed that 25 eV N ion implantation at 5×10^{14} ions/cm^2 fluence resulted in doping of epitaxial graphene (EG) with N concentration of around 1% [14]. Ion implantation on EG with 25 eV B and N ions showed an increase in resistance and a change in magnetoresistance from positive to negative [15]. Defect differences appeared at 200 keV N and Ar ion irradiation of epitaxial graphene, along with a decrease in corrugation [16]. Previously we showed that ion beam implantation (30 keV) and ion beam irradiation (100 MeV) using silver ions acts as a tool to increase sensing capacity of EG on SiC [17,18].

Earlier studies explored defect generation in graphene and SiC independently [19–22]. Only some studies are about defect generation in both graphene and SiC. The substrate plays a crucial role in defect generation in EG on SiC, which shows the importance to study defect generation in both EG and SiC as defect generation in both graphene and SiC enhances their applicability in various interdisciplinary areas [6]. This study presents the effects of ion implantation using N ions on structural properties of EG and SiC. The EG/SiC was irradiated with 10 keV N$^+$ ions at 1×10^{12} ions/cm^2 to 1×10^{14} ions/cm^2 fluences. This range of energy of incident ions create defects both in graphene and SiC while other energy ranges (eV and MeV) influence either graphene or the SiC substrate. Further, doping of graphene with N is advantageous as it results in n-type doping of graphene due to the reason that C and N have comparable atomic radii and chemical bond lengths, which causes substitution of C atom by N atom [23]. Theory revealed that different N dopant configurations result in affecting the local density of states and local charge distribution, and thus changes of the electronic and sensing properties of graphene [24]. The N doping is a way to have n-type graphene material, and is an excellent platform for Li batteries, fuel cells, supercapacitors. This plays a critical role for developing graphene-based nanoelectronics and optronics. Although chemical methods also produce n-type graphene, but they have many disadvantages comparatively to ion implantation like lack of precise control on the concentration of the dopant, contamination and inferred secondary impurities and site selectivity. The ion implantation has revolutionized the semiconductor technology to have a significant economic and societal impact. Integration of ion implantation with EG/SiC will be a further and major step-up into technological progress. Ion implantation of graphene with trivalent and pentavalent impurities presents huge prospect for industrial scale, non-chemistry reliant functionalization and processing of 2D materials.

2. Materials and Methods

Substrates of 4H–SiC (0001) was used for the growth of EG. The graphene was grown inside an inductively heated furnace by thermal decomposition of SiC substate at 2000 °C having an argon pressure at 1 atm [25]. Nitrogen ion (N$^+$) implantation was carried out at Inter University Accelerator Center (IUAC), New Delhi, India employing 30 kV tabletop accelerator. N$^+$ ions with 10 keV energy were used for ion implantation at varying fluences between 1×10^{12} ions/cm^2 to 1×10^{14} ions/cm^2. Stopping and range of ions in matter (SRIM) software was exploited for calculating various parameters like projected range of ions and different types of energy losses [26].

The atomic force microscopy (AFM) by Veeco instruments with 3100 dimensions was done in the tapping mode on both as grown and ion-implanted samples. Raman-AFM WITEC model alpha 300 instrument with excitation laser of 532 nm was used for measuring Raman spectra. An axis Ultra DLD instrument having monochromatic Al (Kα) radiation (hv = 1486.6 eV) was used for doing the XPS measurements. The base pressure in analysis chamber was <1.5×10^{-7} Pa when the XPS spectra were recorded. The C 1s, Si 2p and O 1s core level spectra obtained from XPS further were deconvoluted using Casa XPS software package [27].

3. Results

3.1. AFM

The pristine and nitrogen ion-implanted EG at 1×10^{14} ions/cm^2 fluence is shown in Figure 1a,b which are Atomic Force Microscopy (AFM) phase images obtained at scan areas of 10 μm × 10 μm. The pristine graphene is dominantly covered with monolayer graphene (ML) and bilayer (BL) graphene (darker areas) as depicted in Figure 1a. The thickness of graphene was obtained using optical reflectance mapping. Figure 1b shows changes in the surface morphology following ion implantation which resulted in fragmentation of the ML graphene shown by white arrows. Figure 1c is the phase image and Fig 1d is height image of the sample implanted at the highest fluence (1×10^{14} ions/cm^2) showing presence of dot like structures (shown with black arrows) referred as "nanostructures". Their diameters are estimated from software Nanoscope III (V5.3) (Digital Instruments/Vecco Metrology group, Plainview, NY, USA) to be between 40–70 nm and the heights of around 0.8–1.2 nm.

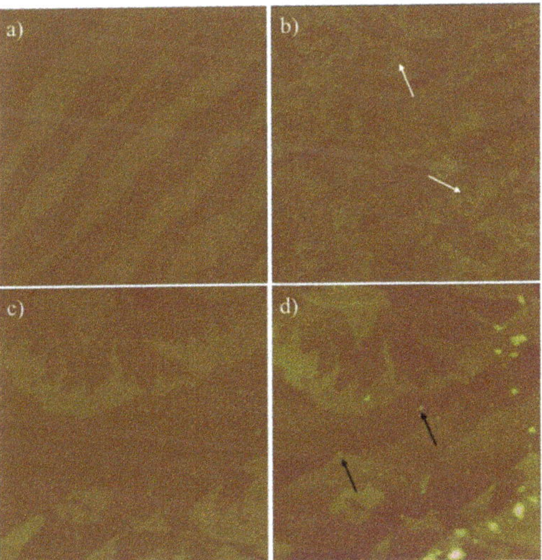

Figure 1. Atomic force microscopy (AFM) phase images of (**a**) pristine and (**b**) N$^+$ ion-implanted epitaxial graphene (EG) on SiC at 1×10^{14} ions/cm^2 fluence over a scan area of 10 μm × 10 μm with white arrows showing fragmentation of Monolayer (ML) graphene; (**c**) phase and (**d**) height image of N$^+$ ion-implanted EG on SiC at 1×10^{14} ions/cm^2 fluence over scan area 2 μm × 2 μm with black arrows showing presence of nanostructures.

At the elastic impact of the energetic ions on the atoms of the material, i.e., in case of nuclear energy loss, the atoms of the material are dislocated from their initial sites in the lattice when the transmitted energy is higher than the displacement threshold energy. The minimum energy required for displacing a single C atom in graphene is 22.2 eV [28]. The minimum lattice displacement energy for the C atom is 20 eV and the Si atom is 35 eV in SiC [29]. In this case, the energy of the incident energetic ions (nitrogen ions) is much more than the energy required for displacing C atoms in graphene and Si and C atoms in the SiC lattice. This resulted in creation of C and Si Frenkel defects. Further, if the surface binding energy required by the atom to bind to the solid surface is lower than the incident energy of the ions, then the atom can be sputtered. In SiC, the surface binding energy was reported to be 7.4 eV for C and 4.7 eV for Si [30].

As the transferred energy is greater than the surface binding energy and the displacement threshold energy of Si and C atoms, it likely results in both displacement and sputtering of target atoms. We estimated the electronic energy loss (Se) and the nuclear energy loss (Sn) by the SRIM software for N^+ ion implantation in EG on SiC, and the results are tabulated in Table 1. It was found that in N^+ ion implantation, Sn dominates as compared to Se.

Table 1. Electronic and nuclear energy loss, longitudinal straggling and lateral straggling calculated from stopping and range of ions in matter (SRIM) simulations.

Ion Type	Ion Energy (keV)	Electronic Energy Loss (S_e) (eV/nm)	Nuclear Energy Loss (S_n) (eV/nm)	Projected Range (nm)	Longitudinal Straggling (nm)	Lateral Straggling (nm)	S_n/S_e
N^+	10	0.1013	0.3104	10.4	3.0	1.9	3

The projected range for N^+ ion implantation is 10.4 nm along with a longitudinal straggling of 3 nm. This shows that most of the damages are generated due to the ion implantation are settled not in the graphene layer, but into the SiC substrate. In this work, defects in graphene are recognized as direct and indirect defects. Direct defects are referred to as defects that arise by the direct impact of the incident ions with C atoms of graphene. Indirect defects are the defects that are created by the impact of the substrate atoms on the C atom of graphene. During the process of collision of incident ions with the substrate atoms, the substrate atoms are sputtered. The sputtered substrate atoms further undergo collision with the C atom of graphene and remove the C atom from the graphene lattice. This process of defect generation in graphene is referred to as indirect defects [6]. It was reported previously that sputtered C atoms undergo agglomeration into small clusters due to their high structural fluidity during the process of ion irradiation [31,32]. Direct and indirect defects cause agglomeration of sputtered atoms which presumably resulted in the formation of nanostructures that are observed in this study on ion-implanted EG/SiC in Figure 1c,d. The Monte Carlo simulation of defect generation upon 0.1 to 1000 keV energy range in ML graphene supported by SiO_2 substrate showed that C Frenkel pairs are primary defect than Si Frenkel pairs [6]. Defect generation in 0.25–50 keV ion-implanted SiC using molecular dynamics simulation reported that most defects are carbon vacancies and carbon interstitials with small fraction of antisite defects (atoms of different type exchanged position) [29]. It was proposed that the substrate plays a major role in enhancing the sputtering yield [6] when C sputtering yield was presented to be higher in graphene supported by SiO_2 substrate than in suspended graphene. Sputtering following ion implantation breaks graphitic carbon bonds and causes graphene fragmentation as observed in Figure 1b. AFM results showed that ion implantation provides a single step process for nano-pattering of EG/SiC in controlled manner which can be further used for developing optoelectronic, electronic and magnetronic devices of high performance.

3.2. Raman Spectroscopy

Raman spectroscopy provides structural fingerprints of the molecules in a material and is a nondestructive method. Figure 2 presents spectra obtained from Raman spectroscopy of pristine and 10 keV N^+ ion-implanted EG at varying fluences. The Raman spectra of graphene are characterized by the G peak located around ~1580 cm^{-1} and a 2D peak present at 2730 cm^{-1} wavenumbers. The G peak is the only first order peak and is a result of bond stretching both by aromatic ring atoms and chain atoms [33]. The 2D peak indicates defect free graphene and is related to a two phonons process. D peak which occurs at ~1350 cm^{-1} wavenumber is related to defects in graphene and is not observed for pristine graphene [34]. Additionally, the presence of D+D$'$ at 2950 cm^{-1} and D$'$ at 1620 cm^{-1}, also demonstrate defects in EG [35].

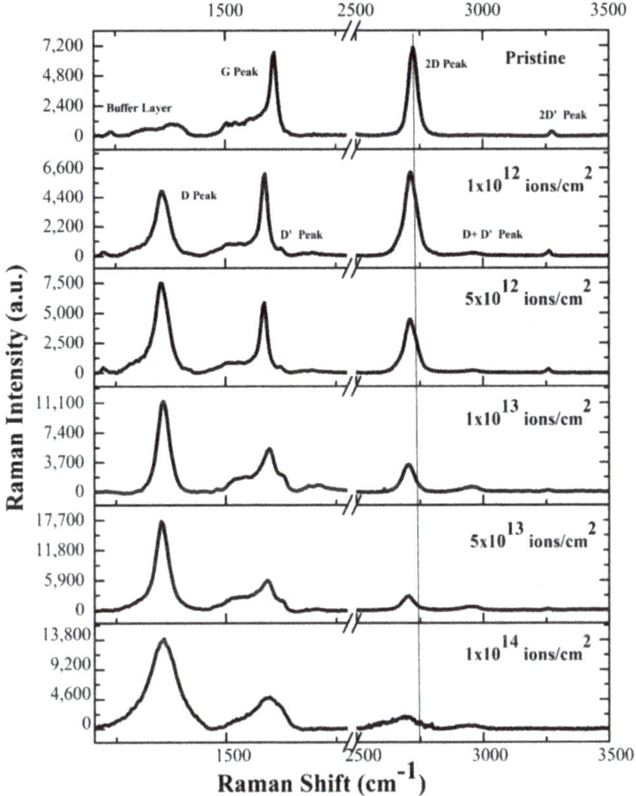

Figure 2. Raman spectra of pristine and N^+ ion-implanted EG on SiC obtained at different fluences.

Figure 2 shows the absence of the D peak in pristine graphene, thus showing as grown graphene is of high quality. Figure 2 shows the presence of the D peak in all ion-implanted samples. The D$'$ peak is clearly visible up to 5×10^{13} ions/cm^2 fluence, however, at the highest fluence (1×10^{14} ions/cm^2) the D$'$ peak is not visible as it merged with the G peak. The D+D$'$ peak appeared in all ion-implanted samples, reflecting a significant amount of defects [35]. The intensity ratio of D peak and G peak is represented as I(D)/I(G) and the intensity ratio of 2D peak and G peak is represented as I(2D)/I(G). They were calculated using Figure 2 for all ion-implanted samples. Defects are generated following ion implantation which resulted in the decrease in I(2D)/I(G) and an increase I(D)/I(G) which was also evident from a previous study [36]. Figure 2 also shows the fluence dependent rise in the intensity of D peak and fall in the intensity of the 2D peak. This signifying increase in defects with the increase in fluence. This is similar to a previous study following ion implantation on graphene sheets. The graphene sheets were prepared by the direct liquid exfoliation of graphite layers which then were deposited on a Si–substrate [37]. There is an increment in the full width half maxima (FWHM) of the 2D peak with increasing fluence.

In graphene, the sp^3 defects dominate when $I(2D)/I(G) > 1$ and $I(D)/I(G) < 1$, while domination by vacancy defects is observed when $I(2D)/I(G) < 1$ and $I(D)/I(G) > 1$ [38]. Another study reported domination of sp^3 defects when $I(D)/I(D')$ is in the range of 7 to 13, and domination of vacancy defects when $I(D)/I(D')$ is in the range of 3.5 to 7 [35]. The authors referred to sp^3 defects as "hopping defects". These defects distort the bonds between the carbon atoms, retaining the general sp^2 configuration, while vacancy defects arise due to missing of atoms from its lattice site. At a fluence of 1×10^{12} ions/cm^2, sp^3 defects dominate as $I(2D)/I(G) > 1$, $I(D)/I(G) < 1$ and $I(D)/I(D')$ is higher than 7 and at a fluence 5×10^{12} ions/cm^2 and higher, vacancy defects dominate as $I(2D)/I(G) < 1$ and $I(D)/I(G) > 1$, and $I(D)/I(D')$ is in the range of 3.5 to 7.

Cancado et al., based on the intensity ratio between D peak and G peak, classified graphene defect generation into two stages, stage I and stage II. In Stage I, there is a rise in the $I(D)/I(G)$ ratio, but in stage II there is a fall in the $I(D)/I(G)$ ratio. The stage II is also marked by a broadening of FWHM of D and 2D peaks [36]. The two stages are discussed in detail by Kaushik et al. for ion implantation using Ag^- ion in epitaxial graphene [17]. Figure 2 shows an increase of the FWHM of the 2D peak in N^+ ion-implanted EG at varying fluences. At fluence of 1×10^{14} ions/cm^2, there is a significant broadening of FWHM of D and 2D peaks. This marks a clear indication of the onset of stage II [36]. Luchesse et al. simulated phenomenological model and gave the relation between $I(D)/I(G)$ and the average distance between defects represented as L_D as shown below [39].

$$\frac{I(D)}{I(G)} = C_A \frac{\left(r_A^2 - r_S^2 \right)}{\left(r_A^2 - 2r_S^2 \right)} \left[e^{-\pi r_S^2/L_D^2} - e^{-\pi(r_A^2-r_S^2)/L_D^2} \right] \tag{1}$$

where r_A = 3.1 nm and r_S = 1 nm are the radii of the activated area and the structurally disordered area. Details of activated and disordered area are mentioned further by Kaushik et al. [17]. The constant C_A = 4.2 which depends on the laser excitation energy (in this work it is 2.33 eV) and the maximum possible value of $I(D)/I(G)$. The $I(D)/I(G)$ were estimated from the results obtained by Raman spectroscopy (Figure 2) and L_D was calculated in graphene using the above equation. The Equation (2) below shows the relation between L_D and defect density represented as n_D.

$$n_D = 10^{14}/\pi L_D^2 \tag{2}$$

The n_D was estimated using Equation (2). Table 2 tabulates the values of $I(D)/I(G)$, L_D and n_D at different fluences in the ion-implanted samples of EG/SiC.

Table 2. Intensity ratios of D and G peaks ($I(D)/I(G)$), mean distance between defects in graphene (L_D) and defect density (n_D) in N^+ ion-implanted EG/SiC.

Samples	N^+ Ion Implantation (10 keV)		
	$I(D)/I(G)$	L_D (nm)	$n_D \times 10^{10}$ (cm^{-2})
Pristine	0.27	22.47	6.31
Fluence			
1×10^{12} ions/cm^2	0.85	10.85	27.05
5×10^{12} ions/cm^2	1.25	8.68	42.26
1×10^{13} ions/cm^2	2.10	5.96	89.65
5×10^{13} ions/cm^2	3.00	3.05	342
1×10^{14} ions/cm^2	2.70	2.85	393

Table 2 shows the rise in the defect density with the increase of fluence. Figure 2 shows that the 2D peak is positioned at ~2730 cm^{-1} in the pristine sample. Besides reduction in the 2D peak intensity with the increasing fluence, we also observed a red shift of the 2D peak (Figure 2) in the ion-implanted samples. The 2D peak is fitted with a Lorentzian curve and we observed that the 2D peak centered at 2730 cm^{-1} in the pristine sample shifted to 2680 cm^{-1} in the highest implanted fluence. This red shift of the 2D peak is because of relaxation of compressive strain following ion implantation as pristine EG is initially under compressive strain. Earlier studies showed that following electron beam implantation a red shift in the 2D peak is due to tensile strain or due to relaxation of compressive strain [40].

The structural modifications in the 4H–SiC substrate was explored using Raman spectroscopy. Defects in graphene appear due to a combination of direct collision of the incident ions with graphene atoms and due to the impact of indirect collision of the substrate atoms [6]. Thus, it is essential to investigate defects generated in SiC as well. Two prominent peaks in SiC Raman spectra are FTO (2/4) at 776 cm^{-1} and LO at 965 cm^{-1} [41]. Figure 3a shows the presence of these prominent SiC peaks in the Raman spectra. Figure 3a indicates a decrement in the intensity of FTO (2/4) and LO with the rise in fluence, which is similar to that reported earlier in ion irradiated SiC [42]. C sub lattices has displacement energy of 20 eV and Si has of 35 eV in the SiC lattice. Since the transmitted energy in this study is higher than the threshold energy required for C and Si atom, the displacement of Si and C atoms from its lattice site occurs and which causes breaking of the SiC bonds. This results in a reduction of the intensity of the FTO (2/4) peak along with the LO peak, which can be observed in Figure 3. At the highest fluence (1×10^{14} ions/cm^2), a shift and narrowing of the FTO (2/4) peak along with the LO peak is observed (Figure 3a). This is in agreement with a previous study, which reported that this could be due to the fact that point defects trap the free carriers. The LO phonon loses the plasmonic coupling as a result of defects causing a shift in the LO peak [43]. Figure 3b illustrates newly formed Raman bands in the range around ~500–550 cm^{-1} at the highest fluence. This is due to the formation of Si–Si bonds in the samples implanted at a fluence 1×10^{14} ions/cm^2 [44]. In addition to that there also occur less intense bands in the region of 650–800 cm^{-1} in the Raman spectra of the sample with the highest fluence. This could be due to the formation of a new vibrational modes of Si–C following the ion implantation. Sorieul and co-workers reported that defects generated in SiC by ion irradiation result in breakdown of Raman selection rules, which leads to generation of Si–Si and C–C and new Si–C vibrational modes [44]. Further, Figure 3b shows a bump in the region 815–950 cm^{-1} which is due to the superposition of three transverse optical phonons which are 2TO (X), 2TO (W) and 2TO (L) of silicon [45].

The relative Raman intensity which is denoted as RRI is the ratio between the average intensity of the first order Raman modes in the ion-implanted sample and that of the pristine sample [46]. Figure 4a shows a decrease in RRI from its maximum value of 1 in the non-implanted sample to the value of 0.30 in the sample implanted at fluence of 1×10^{14} ions/cm^2. Increased amount of defects with increasing fluence is the reason behind the reduction in the value of RRI, which is in agreement with a previous study [46]. The total disorder, which quantifies the damage of the crystallinity of materials is defined as 1-A$_{norm}$. The A$_{norm}$ is the ratio between the total area under the principal first order Raman band in the implanted material and the area under the principal first order Raman band of the crystalline material [47]. The ranges used for estimating the area under principal first order bands (FTO (2/4) and LO) are ~760 cm^{-1}–800 cm^{-1} and 955 cm^{-1}–990 cm^{-1}. Following ion implantation total area under principal first order bands of implanted sample reduces which results in rise in total disorder in implanted samples. The rise in the value of the total disorder with increasing the fluence signifies the decreased crystallinity of the material with increasing fluence [47]. Figure 4b shows the change in the total disorder at varying fluences following the ion implantation, with increment in the value of the total disorder to 0.73 in the highest fluence implanted sample (1×10^{14} ions/cm^2). Raman results suggest that one can introduce different levels of defects into both EG and SiC by just tuning the fluence which forms a platform for developing high performance supercapacitors, sensors, etc. as defects in controlled manner enhance storage capacity and sensing capacity of EG/SiC.

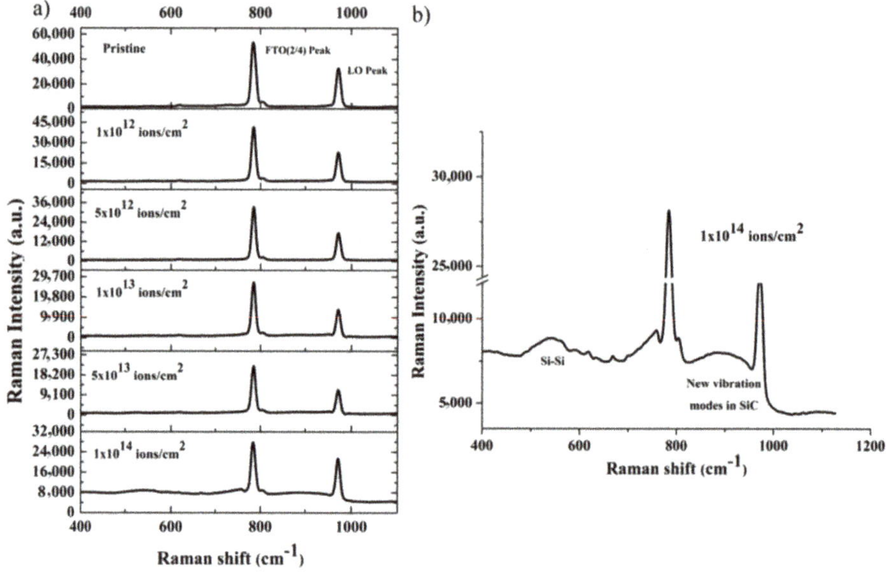

Figure 3. (a) Raman spectra showing changes in the prominent SiC peaks at varying fluences in N$^+$ ion-implanted samples; (b) Raman spectra of the sample implanted at fluence 1×10^{14} ions/cm^2.

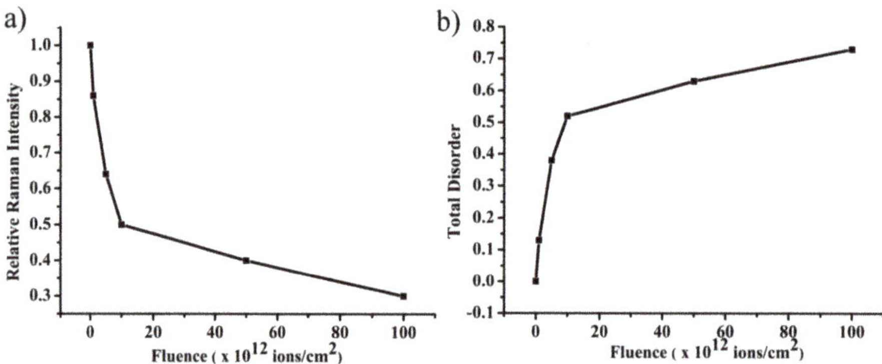

Figure 4. Variation in (a) relative Raman intensity, (b) total disorder following 10 keV N$^+$ ion implantation.

3.3. XPS

Figure 5a shows the C 1s spectra of as grown EG/SiC and implanted EG/SiC at 1×10^{12}, 1×10^{13} and 1×10^{14} ions/cm^2 fluences obtained from XPS at takeoff angle of 90°. The C 1s spectra of EG/SiC are characterized by a graphitic carbon (C=C) peak around 284.8 eV and a silicon carbide (Si–C) peak around ~283.8 eV. Additionally, due to the graphitization process, a minor contribution around 289.0 eV is due to the formation of O-C=O bonds [48] in the pristine sample. Following N$^+$ ion implantation, C 1s spectra show a reduction of the intensity of both graphitic carbon (C=C) and silicon carbide peak (Si–C) than the pristine sample. This is indicative of irradiation-induced damages scaling with an increasing fluence, as revealed by the fact that the overall C 1s and Si 2p spectra intensity decreases when going from 1×10^{12} to 1×10^{14} ions/cm^2.

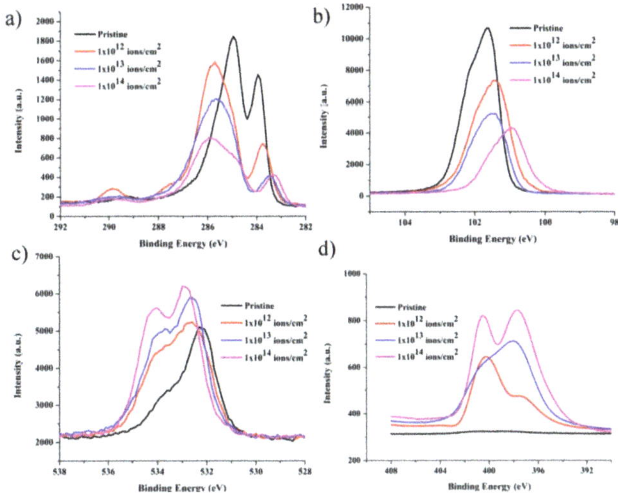

Figure 5. (a) C 1s; (b) Si 2p; (c) O 1s and (d) N 1s spectra obtained from XPS of pristine and N$^+$ ion-implanted EG on SiC samples.

There was a broadening of the C1s spectra in the region between 284.5 eV to 287.5 eV in all ion-implanted samples, Figure 5a It may have been due to the presence of the C=N bond at 285.7 eV following the nitrogen ion implantation in graphene [49]. Figure 5d shows the corresponding N 1s spectra which reveal two components at 397.9 eV and 400.3 eV, that can be assigned to pyridinic and quaternary type N-functionalities, respectively [49,50]. The overall N 1s intensity increased with increasing ion fluence which was indicative of an effective N implantation.

Figure 5b shows Si 2p spectra with the SiC peak at ~101.6 eV in as grown EG/SiC sample in agreement with earlier study [51]. The Si 2p spectra of N$^+$ ion-implanted samples reveal a decreased intensity of the SiC peak with increasing fluence, in agreement with Raman results (Figure 3). In addition, Si 2p spectra shift towards a lower binding energy approaching BE values which is characteristic of pure Si. [52]. These changes can be interpreted as resulting from breakdown of the Si–C bonds and an increased generation of homonuclear bonds of the sp^3 C–C and Si–Si type (Figure 5b).

Figure 5c presents the O 1s spectra which clearly depict a fluence dependent rise in the O 1s peak intensity. There appears two peaks at ~532.1 and 533.2 eV in the O 1s spectra obtained after deconvolution of the O 1s spectra (data not shown) in the as grown EG/SiC sample, peaks are attributed to C=O and C–OH peak, respectively [53]. The high binding energy component increases by the ion implantation which is indicative of C–OH formation. The XPS showed that dopant concentration can be precisely controlled by controlling the fluence which provides a simple method for developing nitrogenated graphene at varying concentrations and enhances the applicability of EG/SiC in nanoelectronics.

4. Conclusions

There was a structural modification of monolayer/bilayer graphene and the SiC substrate using 10 keV N ion implantation. This works shows that at this energy both dopant and defects can be introduced simultaneously in graphene and SiC, while other studies have investigated defect generation in either graphene or SiC. The AFM investigation show a fragmentation of the single and bilayer graphene accompanied by a formation of nanostructures. The ion implantation caused breaking of graphitic carbon bonds which results in the fragmentation of monolayer and bilayer graphene at many places. There was an agglomeration of sputtered carbon atoms into small clusters due to their high structural fluidity which results in formation of nanostructures. The corresponding Raman spectra showed an increase in the D peak intensity up to the fluence of 5×10^{13} ions/cm^2 and a decrease in the D peak intensity at 1×10^{14} ions/cm^2 fluence. Stage I type of defects dominated up to 5×10^{13} ions/cm^2 fluence while stage-II defects were present at the highest fluence. There were sp^3 defects at 1×10^{12} ions/cm^2 fluence and vacancy defects at the higher fluences, as well as an increase in defect density with increase in the fluence. With increasing fluence there was a fall in intensity of the prominent peaks of SiC along with an increase in the value of the total disorder and a decrease in the value of the relative Raman intensity. This was interpreted as the SiC interface with graphene is affected by the N ion bombardment and should be considered when there is device processing. The XPS reveals C=N formation and a fluence dependent decrease in the SiC peak intensity from the Si 2p XPS spectra. A rise in O1s spectra with increase in fluence was observed due to an increase of C=O and C-O-H bond intensity with increasing fluence. Altogether, this work illustrates the possibility that nitrogen doping and different levels of defects may be introduced in both graphene and SiC by N ion implantation on EG/SiC, which implies process control as well as consideration of careful selection of the process conditions for modifying graphene and SiC or both.

Author Contributions: Conceptualization, P.D.K., G.R.Y. and M.S.; formal analysis, P.D.K.; investigation, P.D.K. and G.B.V.S.L.; resources, P.D.K., G.R.Y., G.G., R.Y. and M.S.; visualization, P.D.K.; writing—original draft, P.D.K.; writing—review & editing, P.D.K., G.R.Y., G.B.V.S.L., G.G., R.Y. and M.S.; funding acquisition, P.D.K. and M.S. All authors have read and agreed to the published version of the manuscript.

Funding: This work was supported by research grant from Swedish Research Council (project 2016-06014). P.D. Kaushik acknowledges support from the Åforsk Research Grant (19-675). G.G. acknowledges financial support from VR grant 2018-03957 and Vinnova grant 2019-04882. R.Y. acknowledges financial support via VR grant 2018-04962.

Conflicts of Interest: The authors declare no conflict of interest.

References

1. Geim, A.K.; Novoselov, K.S. The rise of graphene. *Nat. Mater.* **2007**, *6*, 183–191. [CrossRef] [PubMed]
2. Dvorak, M.; Oswald, W.; Wu, Z. Bandgap Opening by Patterning Graphene. *Sci. Rep.* **2013**, *3*. [CrossRef] [PubMed]
3. Dabrowski, P.; Rogala, M.; Wlasny, I.; Klusek, Z.; Kopciuszyński, M.; Jalochowski, M.; Strupiński, W.; Baranowski, J. Nitrogen doped epitaxial graphene on 4H-SiC(0001)—Experimental and theoretical study. *Carbon* **2015**, *94*, 214–223. [CrossRef]
4. Thiele, C.; Felten, A.; Echtermeyer, T.; Ferrari, A.C.; Casiraghi, C.; Löhneysen, H.V.; Krupke, R. Electron-beam-induced direct etching of graphene. *Carbon* **2013**, *64*, 84–91. [CrossRef]
5. Daukiya, L.; Mattioli, C.; Aubel, D.; Hajjar-Garreau, S.; Vonau, F.; Denys, E.; Reiter, G.; Fransson, J.; Perrin, E.; Bocquet, M.-L.; et al. Covalent Functionalization by Cycloaddition Reactions of Pristine Defect-Free Graphene. *ACS Nano* **2017**, *11*, 627–634. [CrossRef] [PubMed]
6. Wu, X.; Zhao, H.; Yan, D.; Pei, J. Investigation on gallium ions impacting monolayer graphene. *AIP Adv.* **2015**, *5*, 067171. [CrossRef]
7. Wang, Q.; Shao, Y.; Ge, D.; Yang, Q.; Ren, N. Surface modification of multilayer graphene using Ga ion irradiation. *J. Appl. Phys.* **2015**, *117*, 165303. [CrossRef]
8. Wang, Q.; Dong, J.; Bai, B.; Xie, G. Investigating change of properties in gallium ion irradiation patterned single-layer graphene. *Phys. Lett. A* **2016**, *380*, 3514–3519. [CrossRef]
9. Kwon, K.C.; Son, P.K.; Kim, S.Y. Ion beam irradiation of few-layer graphene and its application to liquid crystal cells. *Carbon* **2014**, *67*, 352–359. [CrossRef]
10. Xie, G.; Yang, R.; Chen, P.; Zhang, J.; Tian, X.; Wu, S.; Zhao, J.; Cheng, M.; Yang, W.; Wang, D.; et al. A General Route Towards Defect and Pore Engineering in Graphene. *Small* **2014**, *10*, 2280–2284. [CrossRef] [PubMed]
11. Naitou, Y.; Iijima, T.; Ogawa, S. Direct nano-patterning of graphene with helium ion beams. *Appl. Phys. Lett.* **2015**, *106*, 033103. [CrossRef]
12. Morin, A.; Lucot, D.; Ouerghi, A.; Patriarche, G.; Bourhis, E.; Madouri, A.; Ulysse, C.; Pelta, J.; Auvray, L.; Jede, R.; et al. FIB carving of nanopores into suspended graphene films. *Microelectron. Eng.* **2012**, *97*, 311–316. [CrossRef]

13. Xu, Y.; Zhang, K.; Brusewitz, C.; Wu, X.; Hofsass, H.C. Investigation of the effect of low energy ion beam irradiation on mono-layer graphene. *AIP Adv.* **2013**, *3*, 72120. [CrossRef]

14. Willke, P.; Amani, J.A.; Thakur, S.; Weikert, S.; Druga, T.; Maiti, K.; Hofsäss, H.; Wenderoth, M. Short-range ordering of ion-implanted nitrogen atoms in SiC-graphene. *Appl. Phys. Lett.* **2014**, *105*, 111605. [CrossRef]

15. Willke, P.; Amani, J.A.; Sinterhauf, A.; Thakur, S.; Kotzott, T.; Druga, T.; Weikert, S.; Maiti, K.; Hofsäss, H.; Wenderoth, M. Doping of Graphene by Low-Energy Ion Beam Implantation: Structural, Electronic, and Transport Properties. *Nano Lett.* **2015**, *15*, 5110–5115. [CrossRef] [PubMed]

16. Zhao, J.-H.; Qin, X.-F.; Wang, F.-X.; Fu, G.; Wang, X.-L. Raman and morphology visualization in epitaxial graphene on 4H-SiC by Nitrogen or Argon ion irradiation. *Nucl. Instrum. Methods Phys. Res. Sect. B Beam Int. Mater. Atoms* **2015**, *365*, 260–263. [CrossRef]

17. Kaushik, P.D.; Rodner, M.; Lakshmi, G.B.V.S.; Ivanov, I.G.; Greczynski, G.; Palisaitis, J.; Eriksson, J.; Solanki, P.R.; Aziz, A.; Siddiqui, A.M.; et al. Surface functionalization of epitaxial graphene using ion implantation for sensing and optical applications. *Carbon* **2020**, *157*, 169–184. [CrossRef]

18. Kaushik, P.D.; Ivanov, I.G.; Lin, P.-C.; Kaur, G.; Eriksson, J.; Lakshmi, G.; Avasthi, D.; Gupta, V.; Aziz, A.; Siddiqui, A.M.; et al. Surface functionalization of epitaxial graphene on SiC by ion irradiation for gas sensing application. *Appl. Surf. Sci.* **2017**, *403*, 707–716. [CrossRef]

19. Compagnini, G.; Giannazzo, F.; Sonde, S.; Raineri, V.; Rimini, E. Ion irradiation and defect formation in single layer graphene. *Carbon* **2009**, *47*, 3201–3207. [CrossRef]

20. Ochedowski, O.; Bussmann, B.K.; D'Etat, B.B.; Lebius, H.; Schleberger, M. Manipulation of the graphene surface potential by ion irradiation. *Appl. Phys. Lett.* **2013**, *102*, 153103. [CrossRef]

21. Sorieul, S.; Costantini, J.-M.; Gosmain, L.; Calas, G.; Grob, J.-J.; Thomé, L. Study of damage in ion-irradiated α-SiC by optical spectroscopy. *J. Phys. Condens. Matter* **2006**, *18*, 8493–8502. [CrossRef] [PubMed]

22. Kaushik, P.D.; Aziz, A.; Siddiqui, A.M.; Lakshmi, G.; Syväjärvi, M.; Yakimova, R.; Yazdi, G.R. Structural and optical modification in 4H-SiC following 30 keV silver ion irradiation. (International Conference on Inventive Research in Material Science and Technology, Hotel Arcadia, Coimbatore, Tamil Nadu, India, 23–24 March 2018). *AIP Conf. Proc.* **2018**, *1966*, 020035. [CrossRef]

23. Granzier-Nakajima, T.; Fujisawa, K.; Anil, V.; Terrones, M.; Yeh, Y.-T. Controlling Nitrogen Doping in Graphene with Atomic Precision: Synthesis and Characterization. *Nanomater* **2019**, *9*, 425. [CrossRef] [PubMed]

24. Lee, W.J.; Lim, J.; Kim, S.O. Carbon Nanomaterials: Nitrogen Dopants in Carbon Nanomaterials: Defects or a New Opportunity? (Small Methods 1-2/2017). *Small Methods* **2017**, *1*. [CrossRef]

25. Yazdi, G.; Vasiliauskas, R.; Iakimov, T.; Zakharov, A.; Syväjärvi, M.; Yakimova, R. Growth of large area monolayer graphene on 3C-SiC and a comparison with other SiC polytypes. *Carbon* **2013**, *57*, 477–484. [CrossRef]

26. Ziegler, J.F.; Ziegler, M.D.; Biersack, J.P. SRIM—The stopping and range of ions in matter. *Nucl. Instrum. Methods Phys. Res. Sect. B Beam Interact. Mater. At.* **2010**, *268*, 1818–1823. [CrossRef]

27. Walton, J.; Wincott, P.; Fairley, N.; Carrick, A. *Peak Fitting with CasaXPS: A Casa Pocket Book*; Accolyte Science: Knutsford, UK, 2010.

28. Lehtinen, O.; Kotakoski, J.; Krasheninnikov, A.V.; Tolvanen, A.; Nordlund, K.; Keinonen, J. Effects of ion bombardment on a two-dimensional target: Atomistic simulations of graphene irradiation. *Phys. Rev. B* **2010**, *81*, 153401. [CrossRef]

29. Devanathan, R.; Weber, W.J.; Gao, F. Atomic scale simulation of defect production in irradiated 3C-SiC. *J. Appl. Phys.* **2001**, *90*, 2303–2309. [CrossRef]

30. Chang, J.; Cho, J.-Y.; Gil, C.-S.; Lee, W.-J. A Simple method to calculate the displacement damage cross section of silicon carbide. *Nucl. Eng. Technol.* **2014**, *46*, 475–480. [CrossRef]

31. Saito, Y.; Yoshikawa, T.; Inagaki, M.; Tomita, M.; Hayashi, T. Growth and structure of graphitic tubules and polyhedral particles in arc-discharge. *Chem. Phys. Lett.* **1993**, *204*, 277–282. [CrossRef]

32. Dutta, N.; Mohanty, S.; Buzarbaruah, N. Modification on graphite due to helium ion irradiation. *Phys. Lett. A* **2016**, *380*, 2525–2530. [CrossRef]

33. Malard, L.M.; Pimenta, M.A.; Dresselhaus, G.; Dresselhaus, M.S. Raman spectroscopy in graphene. *Phys. Rep.* **2009**, *473*, 51–87. [CrossRef]

34. Beams, R.; Cançado, L.G.; Novotny, L. Raman characterization of defects and dopants in graphene. *J. Phys. Condens. Matter* **2015**, *27*, 83002. [CrossRef] [PubMed]

35. Eckmann, A.; Felten, A.; Mishchenko, A.; Britnell, L.; Krupke, R.; Novoselov, K.S.; Casiraghi, C. Probing the Nature of Defects in Graphene by Raman Spectroscopy. *Nano Lett.* **2012**, *12*, 3925–3930. [CrossRef] [PubMed]

36. Cançado, L.G.; Jorio, A.; Ferreira, E.; Stavale, F.; Achete, C.A.; Capaz, R.B.; Moutinho, M.; Lombardo, A.; Kulmala, T.S.; Ferrari, A.C. Quantifying Defects in Graphene via Raman Spectroscopy at Different Excitation Energies. *Nano Lett.* **2011**, *11*, 3190–3196. [CrossRef] [PubMed]

37. Mishra, M.; Alwarappan, S.; Kanjilal, D.; Mohanty, T. The Effect of Low Energy Nitrogen Ion Implantation on Graphene Nanosheets. *Electron. Mater. Lett.* **2018**, *14*, 488–498. [CrossRef]

38. Zandiatashbar, A.; Lee, G.-H.; An, S.J.; Lee, S.; Mathew, N.; Terrones, M.; Hayashi, T.; Picu, R.; Hone, J.; Koratkar, N. Effect of defects on the intrinsic strength and stiffness of graphene. *Nat. Commun.* **2014**, *5*, 3186. [CrossRef] [PubMed]

39. Lucchese, M.; Stavale, F.; Ferreira, E.; Vilani, C.; Moutinho, M.; Capaz, R.B.; Achete, C.; Jorio, A. Quantifying ion-induced defects and Raman relaxation length in graphene. *Carbon* **2010**, *48*, 1592–1597. [CrossRef]

40. Murakami, K.; Kadowaki, T.; Fujita, J.-I. Damage and strain in single-layer graphene induced by very-low-energy electron-beam irradiation. *Appl. Phys. Lett.* **2013**, *102*, 43111. [CrossRef]

41. Tsukamoto, T.; Hirai, M.; Kusaka, M.; Iwami, M.; Ozawa, T.; Nagamura, T.; Nakata, T. Annealing effect on surfaces of 4H(6H) SiC(0001)Si face. *Appl. Surf. Sci.* **1997**, *113*, 467–471. [CrossRef]

42. Kaushik, P.D.; Aziz, A.; Siddiqui, A.M.; Greczynski, G.; Jafari, M.J.; Lakshmi, G.; Avasthi, D.; Syväjärvi, M.; Yazdi, G.R. Modifications in structural, optical and electrical properties of epitaxial graphene on SiC due to 100 MeV silver ion irradiation. *Mater. Sci. Semicond. Process.* **2018**, *74*, 122–128. [CrossRef]

43. Piluso, N.; Camarda, M.; Anzalone, R.; La Via, F. Micro-Raman Characterization of 4H-SiC Stacking Faults. *Mater. Sci. Forum* **2014**, *778*, 378–381. [CrossRef]

44. Sorieul, S.; Costantini, J.; Gosmain, L.; Thomé, L.; Grob, J.J. Raman Spectroscopy Study of Heavy-Ion-Irradiated Alpha-SiC. *J. Phys. Condens. Matter.* **2006**, *18*, 5235. [CrossRef]

45. Iatsunskyi, I.; Nowaczyk, G.; Jurga, S.; Fedorenko, V.; Pavlenko, M.; Smyntyna, V. One and two-phonon Raman scattering from nanostructured silicon. *Optik* **2015**, *126*, 1650–1655. [CrossRef]

46. Héliou, R.; Brebner, J.; Roorda, S. Optical and structural properties of 6H–SiC implanted with silicon as a function of implantation dose and temperature. *Nucl. Instrum. Methods Phys. Res. Sect. B Beam Interact. Mater. Atoms* **2001**, *175*, 268–273. [CrossRef]

47. Menzel, R.; Gärtner, K.; Wesch, W.; Hobert, H. Damage production in semiconductor materials by a focused Ga+ ion beam. *J. Appl. Phys.* **2000**, *88*, 5658–5661. [CrossRef]

48. Singh, B.; Murad, L.; Laffir, F.; Dickinson, C.; Dempsey, E. Pt based nanocomposites (mono/bi/tri-metallic) decorated using different carbon supports for methanol electro-oxidation in acidic and basic media. *Nanoscale* **2011**, *3*, 3334–3349. [CrossRef] [PubMed]

49. Guo, M.; Li, D.; Zhao, M.; Zhang, Y.; Geng, D.; Lushington, A.; Sun, X. Nitrogen ion implanted graphene as thrombo-protective safer and cytoprotective alternative for biomedical applications. *Carbon* **2013**, *61*, 321–328. [CrossRef]

50. Jiang, Z.-J.; Jiang, Z. Interaction Induced High Catalytic Activities of CoO Nanoparticles Grown on Nitrogen-Doped Hollow Graphene Microspheres for Oxygen Reduction and Evolution Reactions. *Sci. Rep.* **2016**, *6*, 27081. [CrossRef] [PubMed]

51. Riedl, C.; Coletti, C.; Starke, U. Structural and electronic properties of epitaxial graphene on SiC(0 0 0 1): A review of growth, characterization, transfer doping and hydrogen intercalation. *J. Phys. D Appl. Phys.* **2010**, *43*, 374009. [CrossRef]

52. Bashouti, M.Y.; Sardashti, K.; Ristein, J.; Christiansen, S. Early stages of oxide growth in H-terminated silicon nanowires: Determination of kinetic behavior and activation energy. *Phys. Chem. Chem. Phys.* **2012**, *14*, 11877. [CrossRef] [PubMed]

53. Oh, Y.J.; Yoo, J.; Kim, Y.I.; Yoon, J.K.; Na Yoon, H.; Kim, J.-H.; Bin Park, S. Oxygen functional groups and electrochemical capacitive behavior of incompletely reduced graphene oxides as a thin-film electrode of supercapacitor. *Electrochim. Acta* **2014**, *116*, 118–128. [CrossRef]

Article

Electrochemical Deposition of Copper on Epitaxial Graphene

Ivan Shtepliuk [1,2,*], Mikhail Vagin [1,3] and Rositsa Yakimova [1]

[1] Semiconductor Materials, Department of Physics, Chemistry and Biology-IFM, Linköping University, SE-58183 Linköping, Sweden; mikhail.vagin@liu.se (M.V.); rositsa.yakimova@liu.se (R.Y.)
[2] Frantsevich Institute for Problems of Materials Science, NASU—National Academy of Sciences of Ukraine, 142 Kyiv, Ukraine
[3] Laboratory of Organic Electronics, Department of Science and Technology, Linköping University, SE-60174 Norrköping, Sweden
* Correspondence: ivan.shtepliuk@liu.se; Tel.: +46-13282528

Received: 31 January 2020; Accepted: 16 February 2020; Published: 19 February 2020

Abstract: Understanding the mechanism of metal electrodeposition on graphene as the simplest building block of all graphitic materials is important for electrocatalysis and the creation of metal contacts in electronics. The present work investigates copper electrodeposition onto epitaxial graphene on 4H-SiC by experimental and computational techniques. The two subsequent single-electron transfer steps were coherently quantified by electrochemistry and density functional theory (DFT). The kinetic measurements revealed the instantaneous nucleation mechanism of copper (Cu) electrodeposition, controlled by the convergent diffusion of reactant to the limited number of nucleation sites. Cu can freely migrate across the electrode surface. These findings provide fundamental insights into the nature of copper reduction and nucleation mechanisms and can be used as a starting point for performing more sophisticated investigations and developing real applications.

Keywords: epitaxial graphene; copper; redox reaction; electrodeposition; voltammetry; chronoamperometry; DFT

---

1. Introduction

Copper (Cu) deposition is one of the key processes in the electronic industry for circuit interconnections, increasingly replacing aluminum [1]. The electrodeposition of copper has the advantages over physical and chemical vapor deposition methods since it seems to be cheapest and best method to fill vias and trenches. Moreover, copper is one of the most-investigated metal catalysts, stipulating studies of electrodeposition on a variety of carbon materials, especially graphite [2–4].

The electronic conductivity of graphite is established due to the planar conjugation of the sp^2 bonds in a single graphene sheet as a primary building block of all graphitic materials. This defines a significant anisotropy of the density of states (DOS) for in-plane and out-of-plane (edge and basal planes, correspondingly) electronic conduction, which is the origin of the duality in the general properties of all graphitic materials [5,6]. The use of macroscale defect-free graphene sheets [7] allows for the avoidance of contributions from the reactive edge plane of high DOS. In the past decade, huge efforts have been made to conceptualize graphene–copper material hybridization for practical applications [8–13]. Particularly, it was discovered that the addition of graphene to the copper matrix can improve the mechanical and electrical properties of Cu, causing a significant increase in the electrical conductivity, Young's modulus, shear modulus, and Vickers hardness as well as a reduction in the thermal expansion coefficient [8–11]. The aforementioned advantages create excellent prerequisites for copper–graphene (Cu–Gr) composites to be used as reliable interconnection materials, electrical contact materials for ultrahigh-voltage circuit breakers and printed electronics. The enhanced conductivity of the Cu–Gr composite also determines its exploitation as a high-performance counter electrode in solar cells [12,13]. Furthermore, due to the larger heat transfer coefficient and higher corrosion resistance of Cu–Gr hybrids in 3.5% NaCl medium when compared to pure copper [14–18], these materials are very promising for engineering applications (for instance, anti-corrosion coatings) in sea water and for heat transfer applications, respectively. Copper–graphene nanohybrids were, however, recognized and appreciated mainly for their role in the development of efficient catalytic and ultrasensitive detection technologies. The benefits gained from the interaction between graphene-based materials and copper have been used for reduction of 4-nitrophenol (4-NP) to 4-aminophenol [19,20], oxidation of hydrazine [21,22], electro-oxidation of methanol [23], hydrogen evolution reaction (HER) [24,25], the electrochemical reduction of CO_2 to ethanol [26], oxidative carbonylation of methanol [27], CO_2 cycloaddition to propylene oxide (PO) [28], formic acid synthesis by CO_2 hydrogenation [29] and CO_2 electroreduction for methane and methanol production [30]. Beyond the issue of high-performance catalysts, it is worth mentioning that Cu–Gr composites have also been applied to the electrochemical detection of ascorbic acid and dopamine [31], organophosphorus pesticide [32], heavy metals [33,34], glucose [35–39], chlorophenol pollutants in wastewater [40], hydroquinone and catechol [41], nitrite [42,43], nitrogen dioxide (NO_2) [44] and hydrogen peroxide (H_2O_2) [45]. Numerous density functional theory (DFT) calculations shed light on the adsorption of different gases (H_2S [46,47], CO [48], CO_2 [49], N_2O [50], H_2 [51–53]) and organic molecules [54–56] onto Cu-decorated/doped/anchored graphene, thereby providing the solid theoretical background that is required to design efficient sensing devices.

Until now we have been giving an overview of most relevant applications of copper–graphene composite materials. However, there is another aspect of mutual interplay between graphene and copper that needs to be addressed: the detection and removal of Cu^{2+} ions in/from potable water. Even though Cu is the third most abundant transition metal in the human body and it is vitally important for human health, an excess of copper ions can cause cirrhosis of the liver in children, Wilson's disease, Alzheimer's disease, etc. [57,58]. For this reason, the development of ultra-sensitive and selective methods for the monitoring and determination of trace Cu^{2+} ions in water is highly demanded. Since graphene enables the detection of single molecules accommodated on its surface [59], it attracts a lot of attention as a highly sensitive and biofriendly material for environmental sensorics [60], including the fast and real-time detection of Copper ions [61]. Two strategies in such a direction have been implemented so far: (i) electrochemical sensing by electrodes covered with functionalized graphene-based materials (reduced graphene oxide, graphene oxide, graphene quantum dots) [62–65] and (ii) optical sensing using colorimetric and fluorescent probes based on graphene derivatives [66–74]. Both approaches present some limitations, mainly linked to the instability and complicated preparation procedure of the working electrode or fluorescence probe, which typically involves surface functionalization and multistage chemical reactions. In most cases, such a functionalization can increase the toxicity of a sensitive component and is, therefore, less desired.

Exploiting epitaxial graphene grown on SiC (epitaxial graphene (EG)/SiC) using the high-temperature sublimation technique is regarded as an alternative strategy for sensing platform development. Due to evident advantages over other graphene family materials (namely, its large surface area free of functional groups, high quality of monolayer graphene, thickness uniformity, wide potential window, high signal-to-noise ratio, transfer-free technology, and direct sublimation growth without precursors [75–77]), epitaxial graphene has been tested with promising results for the real-time detection of Pb^{2+} ions in aqueous solutions (with a detection limit far below the WHO's permissible level for lead in drinking water) through performing square-wave anodic stripping voltammetry and response–recovery measurements [78,79]. Inspired by this, we have recently extended our activities beyond lead detection to include the investigation of the electrochemical behavior of the Hg^{2+}/Hg^0 redox couple at the epitaxial graphene working electrode [80]; now, we are aiming to elucidate the nature of the copper electroreduction at the epitaxial graphene surface. Since this issue is very poorly investigated in the literature, in-depth investigations at an atomistic level are still needed to deal with the problem. In this regard, complementary electrochemical and theoretical studies of the graphene–Cu system presented here may provide critical information on the Cu kinetics, diffusion, adsorption and sensing mechanisms, thereby facilitating the development of a real-time detection platform. Leaving aside the Cu^{2+} detection by using graphene, it should be recalled that a holistic understanding of the interplay between epitaxial graphene and copper is essential not only for sensing applications, but also for catalytic and electrical applications, due to the mutually reinforcing character of the graphene–Cu pair.

In this report, the electrodeposition of copper on epitaxial graphene has been investigated in acidic environment. The electrodeposition of copper occurs by instantaneous nucleation at a limited number of sites, via a bi-electronic reduction bypassing of Cu^{1+} intermediates at a significant overpotential with respect to the process in graphitic materials. The computational methods showed coherence with the experimental observations.

2. Materials and Methods

2.1. Reagents

Copper (II) sulfate and perchloric acid were purchased from Sigma (Stockholm, Sweden) and used as received. Experiments were carried out with Milli-Q water from a Millipore Milli-Q system.

2.2. Samples and Processes

An Autolab type III potentiostat (Autolab, EcoChemie, Utrecht, Netherlands) was used for the electrochemical measurements. An Ag/AgCl electrode in 3 M KCl and a platinum wire were employed as reference and counter electrodes, respectively, for all measurements.

The samples of nominally monolayer epitaxial graphene (EG) on SiC (substrate area 7×7 mm^2 and thickness 0.4 mm) were obtained from Graphensic AB and produced by the high temperature thermal decomposition of the Si-face (0001) 4H-SiC substrate (7×7 mm^2) in an argon atmosphere using an inductively heated graphite container with a well-controlled temperature profile [81]. Most of the substrate surface is coated with monolayer graphene after graphenization, while the fraction of bilayer graphene islands is negligibly small, which is evidenced by optical reflectance mapping [82]. The samples of epitaxial graphene were used as a working electrode in the open electrochemical cell obtained from Redoxme AB [83]. The cell consisted of a cup of 300 μL with a 2 mm diameter hole and a Vitron o-ring on the bottom. The EG sample was fixed under the hole with the o-ring using screws on the lid. A dry contact for the EG was formed by an aluminum adhesive. The mounted EG sample was kept inside the cell during all wet measurements and procedures in order to avoid sample drift.

The electrolyte solution was prepared from ultrapure 0.1 mol·L^{-1} HClO$_4$, 0.1 mM Cu^{2+} (the purity of copper sulfate, CuSO$_4$, was higher than 99%), and Milli-Q water. The non-complexing character of the perchlorate ions (ClO$_4^-$), with respect to metal cations in aqueous solutions [84–86], suggests that the cyclic voltammetry (CV) measurements will enable us to gain insights into Cu-involved oxidation–reduction reactions, rather than reactions involving more complicated chemical complexes. To explore the Cu redox behavior and kinetics at the working electrode, we exploited the quite concentrated aqueous solution of Cu salt to ensure an intense electrochemical signal. The scan rate was 20 mV/s. The electrochemical reactions are expected to occur at the nominal electrode area of 3.1 mm^2. To elucidate the nature of the Cu kinetics, current–time transients were recorded during the early stages of Cu electrodeposition on Gr/SiC.

2.3. Computational Methods

The Scharifker–Hills approach [87] has been applied to shed light on the nucleation mechanism. The electrochemistry of Cu at the graphene surface was also investigated through in-depth functional theory (DFT) calculations performed by using the Gaussian 16 Rev. B.01 package [88]. The $C_{150}H_{30}$ structure was chosen as a model of graphene. Since, in aqueous solution, the water molecules demonstrate a strong tendency to attach to the metal cation and to form the first coordination sphere (primary solvation shell), in the present paper, we consider the interaction of the $[Cu(H_2O)_6]^{2+}$ complexes with graphene, and their reduction to $[Cu(H_2O)_6]^0$ species. All calculations were carried out using the PBE1PBE level of theory, with empirical dispersion corrections [89,90]. The 6–31G(d,p) basis set was used for carbon, oxygen, and hydrogen atoms, while the basis set developed by the Stuttgart–Dresden–Bonn group (SDD) was utilized for the Cu [91]. Noncovalent interaction (NCI) analysis was performed using the Multiwfn program to better understand the Cu–carbon bonding characteristics [92].

3. Results and Discussion

The voltammetry of epitaxial graphene (Figure 1) in the pure supporting electrolyte does not show any faradaic current in the double layer region of potentials (between +0.5 and −0.4 V). The larger negative potentials (below −0.6 V) applied on epitaxial graphene in pure acidic electrolytes were characterized by the appearance of faradaic currents due to the appearance of a hydrogen evolution reaction. The addition of copper ions to the thousand-times excess of background electrolytes led to the increase in capacitive currents (ca. 2.5 times at 0.5 V). This manifests the strong adsorption of di-cations on epitaxial graphene [77,78,93]. In parallel, the redox phenomena associated with the electrodeposition and dissolution of the metal species became explicit. The voltammetry scan in the negative direction shows the appearance of an intensive cathodic process at potentials below −0.2 V. Being a poor complexing ion, perchlorate disables the stabilization and appearance of Cu^{1+}-based intermediates, which is illustrated by the absence of any redox processes at the higher potentials [94,95]. This implies that the copper electrodeposition on epitaxial graphene proceeds via single, irreversible bi-electronic reduction. The nucleation loop observed for the cathodic process is assigned to the new phase formation [96]. Taking into account the differences in the concentration of copper ions as well as the irreversible (non-Nernstian) character of the deposition process [3], the current density of the nucleation loop reaches values comparable with copper electrodeposition on glassy carbon [3] and pyrolytic graphite [4] at the potentials shifted on ca. 400 mV in a negative direction. This overpotential illustrates the typically slower kinetics of the generalized faradaic process on an epitaxial graphene monolayer in comparison with the common bulk graphitic interfaces [97] contributed by reactive graphene edges.

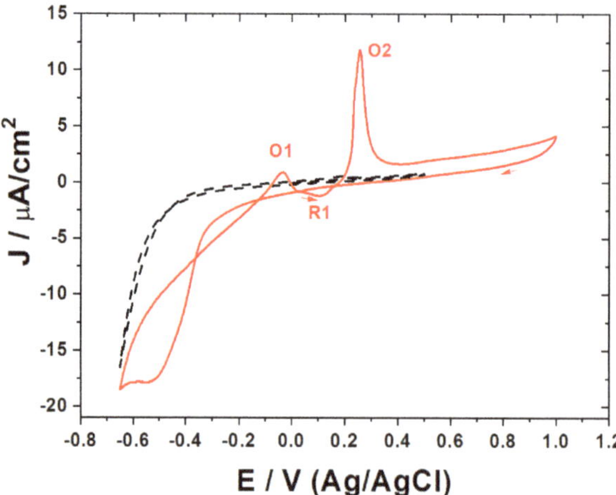

Figure 1. Cyclic voltammograms related to the electrodeposition and stripping of copper on the epitaxial graphene (in absence and in presence 0.1 mM Cu^{2+}, as dashed black and solid red curves, respectively; 0.1 M $HClO_4$, scan rate 20 mV·s^{-1}).

The voltammetry scan in the positive direction showed the appearance of discrete steps of dissolution (stripping) of the copper deposits. The direction-dependent voltammetry response illustrates the difference between the pristine and copper-modified epitaxial graphene monolayer presented at potentials higher than 0.5 and lower than −0.5 V, respectively. The oxidation peak current (O1), located at −0.03 V. corresponds to the single-electron oxidation of metallic copper to Cu^{1+}. Further increases in the applied potential in a positive direction led to the unusual appearance of the negative current (R1) at 0.1 V. In contrast to the kinetically slow electrochemical deposition of copper on the epitaxial graphene monolayer, the aforementioned instability of Cu^{1+} might enable an exergonic (non-faradaic) route towards the metallic copper at the surface via fast disproportionation. Here, the state of the surface is not pristine yet. As soon as the copper ions undergo the faster reduction in metallic copper in comparison with bare carbon [96], the metallic copper might act as an electrocatalyst for the reduction, in comparison with epitaxial graphene originating the appearance of the negative currents. Further increases in potential in a positive direction showed the appearance of the sharp peak current typical in complete oxidation to Cu^{2+} ions. The scenario of the electrode processes was repeated with continuous voltammetry cycling.

We performed DFT calculations in order to assay the reduction thermodynamics. The Cu^{2+} ion was considered, when coordinated with six water molecules [98], energetically, the most favorable complex (Figure 2a). All calculations were performed using the aquo complexes of copper species, because the intrinsic or extrinsic solvent models did not give satisfactory results.

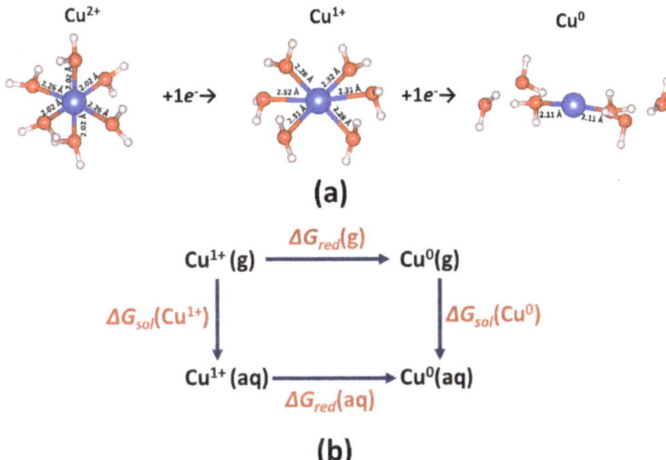

Figure 2. (**a**) Structures of the copper (Cu) complexes: $[Cu(H_2O)_6]^{2+}$, $[Cu(H_2O)_6]^{1+}$, $[Cu(H_2O)_6]^0$. The numbers are bond lengths in Å. Light pink, blue and red spheres represent H, Cu and O atoms, respectively. (**b**) The Born–Haber cycle for the calculation of the potential reductions in $[Cu(H_2O)_6]^{1+}$ to $[Cu(H_2O)_6]^0$.

The sequential additions of water molecules yielded the stabilization of all considered particles, computed as a decrease in the electron acceptor reactivity (Figure S1, Supporting Information) and the Hirshfeld charge on the Cu species (Figure S2, Supporting Information). The aquo complex has disordered octahedral geometry (Jahn–Teller distortion) with four short (2.02 Å) and two elongated Cu–O bonds (2.26 Å). According to the Hirshfeld charge population analysis, the charge located on the copper atom is equal to $+0.601e^-$. The addition of a first electron yielded the aquo Cu^+ complex with a smaller Hirshfeld charge on the copper atom (+0.216) and all elongated Cu–O bonds. The addition of the second electron led to the formation of metallic copper, confirmed by the negligibly small Hirshfeld charge on the copper atom ($+0.005e^-$) and the destruction of the octahedral geometry.

The equilibrium potential of the mono-electronic redox couple $[Cu(H_2O)_6]^{1+}/[Cu(H_2O)_6]^0$, estimated by applying the Born–Haber methodology (Figure 2b) [99], was −0.59 V, which agrees well with the experimental position of the metallic copper deposition. On the contrary, the computed value of the equilibrium potential of the bi-electronic redox couple $[Cu(H_2O)_6]^{2+}/[Cu(H_2O)_6]^0$ was −0.93 V, which is too low in comparison with the observed potential. This might imply that the copper deposition proceeds via mono-electronic reduction, followed by the fast exergonic disproportionation of unstable Cu^{1+} intermediates.

The irreversible process of copper deposition was quantified by the dynamic electrochemical measurements employing Scharifker–Hills formalism [87]. Current transients recorded at different deposition potentials (Figure 3a) are contingent on the applied potential pulse. The short-elapsed times (ca. 0.2 s) are characterized with the equilibration of the electrical double layer at the epitaxial graphene–electrolyte interface. The longer times are characterized with a faradaic process of copper electrodeposition (Inset in Figure 3a). Specifically, the potential pulses of −0.39 and −0.4 V are featured with non-monotonous current transients indicative of the three-dimensional (3D) nucleation process with diffusion-controlled growth [100] which enable analysis of the nucleation mechanism at the early stages of the electrodeposition [87]. The behavior of the active sites available for the formation and growth of the metal nuclei differentiates the two distinctive cases of the nucleation process. Accordingly, the instantaneous nucleation implies the maintenance of the convergent diffusion of the reactant on the limited number of nuclei. On the contrary, progressive nucleation relies on the increase in the number of nuclei yielding the quick establishment of overlapped planar reactant diffusion. The theoretical current transients obey the following equations:

$$\begin{cases} \left(\dfrac{j_{inst}}{j_{max}}\right)^2 = 1.9542\left(\dfrac{t}{t_{max}}\right)^{-1}\left\{1 - \exp\left[-1.2564\left(\dfrac{t}{t_{max}}\right)\right]\right\}^2 \\ \left(\dfrac{j_{prog}}{j_{max}}\right)^2 = 1.2254\left(\dfrac{t}{t_{max}}\right)^{-1}\left\{1 - \exp\left[-2.3367\left(\dfrac{t}{t_{max}}\right)^2\right]\right\}^2 \end{cases} \tag{1}$$

where t and t_{max} are the elapsed and maximum times, respectively; j_{inst}, j_{prog} and j_{max} are the current densities for instantaneous and progressive nucleation, and maximum current density, respectively. The initial kinetics of the copper electrodeposition obeys the three-dimensional instantaneous nucleation model (Figure 3b), which is likely the general mechanism for the metal's electrodeposition on epitaxial graphene monolayer, since the same growth kinetics was observed during electrodeposition of lead [101] and mercury [80]. This implies that metal electrodeposition on epitaxial graphene happens on a limited number of active sites.

The diffusion coefficient of copper ions as a reactant can be estimated from the current transients using the following relationship [102]:

$$D = \frac{j_{max}^2\, t_{max}}{0.1629\cdot(z\cdot F\cdot C)^2} \tag{2}$$

where the t_{max} is maximum time, corresponding to the maximum current j_{max}, z is the valency of the metal ion (+2 in the case of divalent species), F is the Faraday constant (96 485 C/mol), C is the reactant concentration (10^{-7} mol/cm^3). The values of the diffusion coefficient were 4.97×10^{-5} cm^2/s and 5.32×10^{-5} cm^2/s (at −0.39 and −0.40 V, respectively), which is higher than reported for glassy carbon ($\sim$0.4–0.8 $\times 10^{-5}$ cm^2/s) [3]. The nuclei population density on epitaxial graphene monolayer was determined by using the following formula:

$$N_0 = 0.065\left(\frac{1}{8\pi C V_m}\right)^{1/2}\left(\frac{nFC}{I_{max}t_{max}}\right)^2 \tag{3}$$

where n is the number of electrons involved, V_m is the molar volume. N_0 was estimated to be as high as a 1.55×10^6 cm^{-2} and 1.38×10^6 cm^{-2} (at −0.39 and −0.40 V, respectively), which is comparable with data from the literature [3].

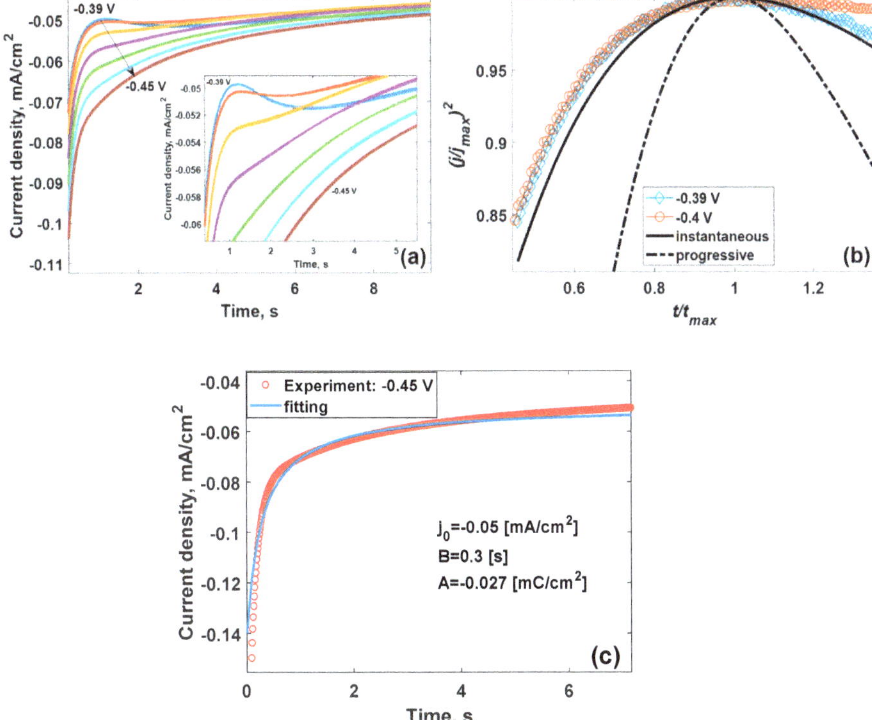

Figure 3. (**a**) Potentiostatic current transients recorded on epitaxial graphene recorded at the cathodic potential pulses of different amplitudes (resting potential 0.5 V; 0.1 M HClO$_4$, 0.1 mM Cu^{2+}). Insert: zoomed current transients; (**b**) comparison of the dimensionless experimental (open symbols) and theoretical transients for instantaneous (bold solid black curve) and progressive (bold dashed–dotted curve black curve) nucleation; (**c**) experimental current transient recorded at −0.45 V (red circles) and Elovich fitting curve (blue solid curve), respectively.

The change of the deposition potential towards more negative values led to the appearance of monotonous current transients (Figure 3a) assigned to the loss of the diffusion control and the appearance of chemisorption phenomena due to the double-layer effect at the electrode surface [103], which can be well described by the modified Elovich equation [103]:

$$J = \frac{A}{B+t} + j_L \tag{4}$$

where A, B and j_L are fitting parameters. It was revealed that that the best fit can be achieved by using the following parameters as follows: $A = -0.027$ mC/cm^2, $B = 0.3$ s and $j_L = -0.05$ mA/cm^2 (Figure 3c).

To better understand the nature of the interaction between electro-reduced species with graphene under realistic conditions, we then simulated the two-stage reduction process by (i) adding one electron directly to the Cu^{2+} ion in the [Cu(H$_2$O)$_6$]$^{2+}$ complex adsorbed on graphene to form [Cu(H$_2$O)$_6$]$^{1+}$ and (ii) adding one more electron to the resulting [Cu(H$_2$O)$_6$]$^{1+}$ complex in order to form the neutral complex [Cu(H$_2$O)$_6$]^0. The result of the structural optimization is shown in Figure 4a. As a direct consequence of the reduction process, the six-coordinated water complex is broken, and all complex's components become weakly bonded to each other. Having a positive charge (according to Hirshfeld population analysis) of +0.29e^-, the Cu atom tends to occupy the top site (directly above the C atom) of the graphene, with an adsorption height of 2.7 Å. The results of the energy decomposition analysis, based on forcefield (EDA-FF), presented in Table 1 showed that the interaction between copper and graphene was very weak and was dominated by dispersion forces, while a strong repulsion is expected between the water molecules and the Cu atom because these two fragments both have a positive charge.

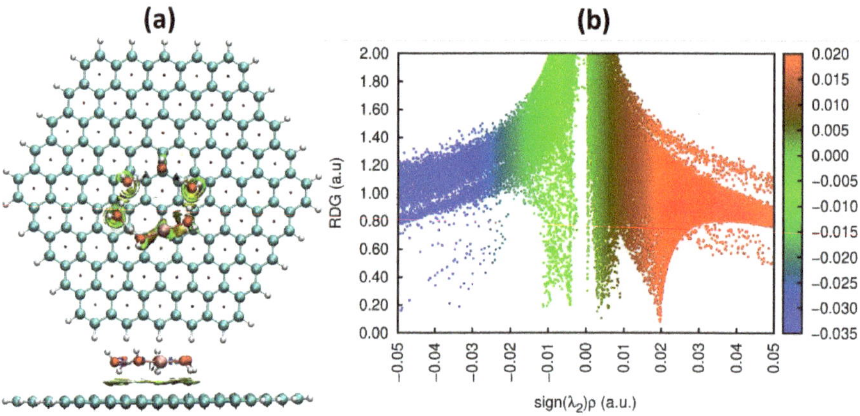

Figure 4. (**a**) Top and side view plots of the noncovalent interaction (NCI) iso-surface (reduced density gradient, RDG = 0.5) for reduced $[Cu(H_2O)_6]^0$ complex on graphene and (**b**) corresponding NCI diagram (RDG vs. $sign(\lambda_2)\rho$). The iso-surfaces are colored according to $sign(\lambda_2)\rho$ in the range −0.035–0.02 a.u. Red indicates the steric repulsion region; green (light brown) indicates the van der Waals interaction region, and blue implies the strong attractive interaction. Note: the optimization of the $[Cu(H_2O)_6]^0$ complex on graphene has been performed with consideration of solvent effect (water in our case) by using polarizable continuum model.

Table 1. Results of energy decomposition analysis.

Interaction between Fragments	Electrostatic, kJ/mol	Repulsion, kJ/mol	Dispersion, kJ/mol	Total Energy, kJ/mol
Graphene–Cu	0.00	1.33	−2.59	−1.26
Graphene–water	−3.58	42.38	−86.01	−47.21
Cu–water	0.00	420.21	−20.10	400.11

Noncovalent interaction (NCI) analysis enabled us to visualize the weak interaction between $[Cu(H_2O)_6]^0$ and graphene, which manifests itself as the green and light brown-colored nonbonding interaction areas located between copper and graphene in the vicinity of the top site (Figure 4a) and in the presence of two spikes within the $sign(\lambda_2)\rho$ region ranging from −0.01 to 0.01 a.u. (Figure 4b).

As was estimated by our DFT calculations, whenever the electro-reduced Cu species reach the graphene surface, they can freely diffuse along both considered diffusion paths with a small energy barrier of 35 meV (Figure 5a,b). Copper atoms tend to avoid occupation of the unfavorable hollow sites. It is interesting to note that the presence of the second copper atom (the already reduced and adsorbed Cu species) at the graphene surface significantly modifies the diffusion path due to the attractive interatomic interaction between neutral Cu species, making the Cu migration energetically favorable and promoting Cu clusterization. From a practical point of view, this means that, due to the favorable mass transfer and nucleation rate, the discrete diffusion zones of individual Cu nuclei will overlap very fast with time. For this reason, the cathodic current reaches a peak current maximum during the first two seconds and then decays slowly (as was demonstrated in Figure 3a).

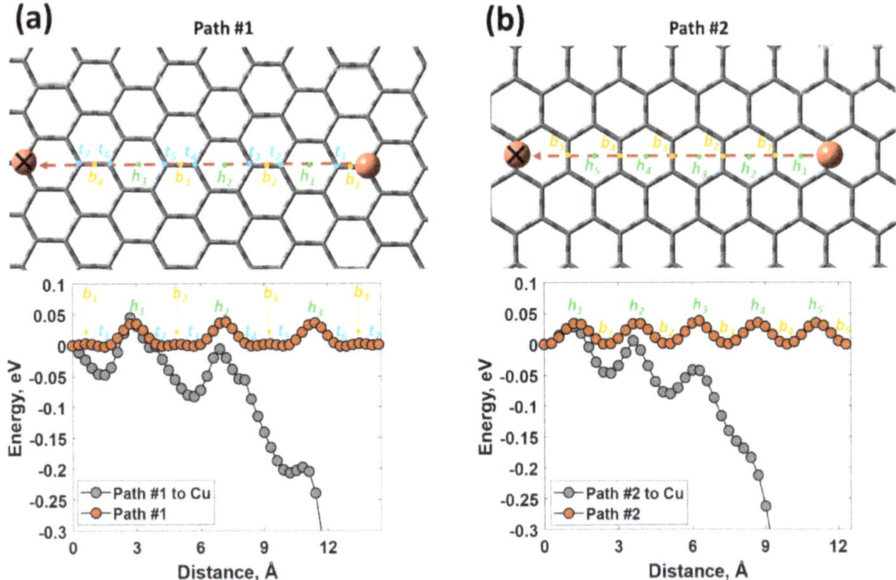

Figure 5. Potential energy curves for the movement of metallic copper atoms along the path #1 (**a**) and path #2 (**b**) in the presence and in the absence (marked by cross) of another copper atom. The height of the Cu atom was set to 2.723 Å. Note: the potential energy curve was normalized to the total energy of the system with a copper atom located at the top site of the graphene (energetically the most favorable site for copper adsorption); t_i, b_i and h_i designate the possible sites (top, bridge and hollow, respectively) that are available for copper atom adsorption.

4. Conclusions

In this work, we have clarified the fundamental mechanisms behind the Cu electroreduction and kinetics at the epitaxial graphene/4H-SiC (0001) working electrode. We demonstrated and discussed the results of cyclic voltammetry, chronoamperometry, and DFT calculations in order to unravel the behavior of copper. The significant overpotential, typical of general faradaic phenomena, on epitaxial graphene yielded a bi-electronic reduction in copper ions bypassing Cu^{1+} intermediates. The dynamic electrochemical measurements revealed an instantaneous nucleation mechanism, implying a limited number of active sites available for the deposit's growth, which seems to be akin to other metal (e.g., Pb, Hg and Li) electrodeposition on epitaxial graphene. In particular, it was revealed that the electrodeposition of metallic copper was possible only at two potentials (−0.39 and −0.4 V), as evidenced by the shape of the corresponding current transients. The estimated diffusion coefficient (~4.97–5.32 × 10^{-5} cm^2/s) from the electrolyte to the EG electrode and the nuclei population density (~1.38–1.55 × 10^6 cm^{-2}) is consistent with data from the literature and indicates that the epitaxial graphene electrode is suitable for fast and Cu electroplating. By performing DFT calculations, we modelled the diffusion of the Cu on graphene and showed that, independently of the diffusion paths, there is a small barrier (35 meV) for the surface migration of Cu, which disappears in the presence of another Cu atom in close proximity to the first one. This favors the mass transfer within the diffusion zones and a fast nucleation process. The obtained results shed light on the nature of the copper electroreduction process at the epitaxial graphene and may facilitate the development of real applications based on copper–graphene nanohybrid materials.

Supplementary Materials: The following are available online at http://www.mdpi.com/2076-3417/10/4/1405/s1, Figure S1: The decrease of the computed electron acceptor reactivity with the addition of the water molecules into the aquo complexes. Figure S2: The change in Hirshfeld charge on Cu species with the addition of the water molecules into the aquo complexes.

Author Contributions: Conceptualization, I.S. and M.V.; methodology, I.S. and M.V.; software, I.S.; validation, I.S., M.V. and R.Y.; formal analysis, I.S.; investigation, I.S. and M.V.; resources, I.S., M.V. and R.Y.; data curation, I.S.; writing—original draft preparation, I.S.; writing—review and editing, I.S., M.V. and R.Y.; visualization, I.S.; supervision, R.Y.; project administration, R.Y.; funding acquisition, I.S. and R.Y. All authors have read and agreed to the published version of the manuscript.

Funding: The authors would like to thank the Swedish Foundation for Strategic research (SSF) for financial support through the grants GMT14-0077 and RMA15-024. I.S. acknowledges the support from Ångpanneföreningens Forskningsstiftelse (Grant 16-541). R.S. acknowledges financial support via VR grant 2018-04962.

Conflicts of Interest: The authors declare no conflict of interest.

References

1. Hu, C.-K.; Harper, J.M.E. Copper interconnections and reliability. *Mater. Chem. Phys.* **1998**, *52*, 5–16. [CrossRef]
2. Liu, H.; Favier, F.; Ng, K.; Zach, M.P.; Penner, R.M. Size-selective electrodeposition of meso-scale metal particles: A general method. *Electrochim. Acta* **2001**, *47*, 671. [CrossRef]
3. Grujicic, D.; Pesic, B. Electrodeposition of copper: The nucleation mechanisms. *Electrochim. Acta* **2002**, *47*, 2901. [CrossRef]
4. Ghodbane, O.; Roue, L.; Belanger, D. Copper electrodeposition on pyrolytic graphite electrodes: Effect of the copper salt on the electrodeposition process. *Electrochim. Acta* **2007**, *52*, 5843. [CrossRef]
5. Brownson, D.A.C.; Kampouris, D.K.; Banks, C.E. Graphene electrochemistry: Fundamental concepts through to prominent applications. *Chem. Soc. Rev.* **2012**, *41*, 6944. [CrossRef] [PubMed]
6. McCreery, R.L. Advanced carbon electrode materials for molecular electrochemistry. *Chem. Rev.* **2008**, *108*, 2646–2687. [CrossRef] [PubMed]
7. Virojanadara, C.; Syväjarvi, M.; Yakimova, R.; Johansson, L.I.; Zakharov, A.A.; Balasubramanian, T. Homogeneous large-area graphene layer growth on 6H-SiC (0001). *Phys. Rev. B* **2008**, *78*, 245403. [CrossRef]
8. An, Z.; Li, J.; Kikuchi, A.; Wang, Z.; Jiang, Y.; Ono, T. Mechanically strengthened graphene–Cu composite with reduced thermal expansion towards interconnect applications. *Microsyst. Nanoeng.* **2019**, *5*, 20. [CrossRef]
9. Cui, R.; Han, Y.; Zhu, Z.; Chen, B.; Ding, Y.; Zhang, Q.; Wang, Q.; Ma, G.; Pei, F.; Ye, Z. Investigation of the structure and properties of electrodeposited Cu/graphene composite coatings for the electrical contact materials of an ultrahigh voltage circuit breaker. *J. Alloy. Compd.* **2019**, *777*, 1159–1167. [CrossRef]
10. Maharana, H.S.; Rai, P.K.; Basu, A. Surface-mechanical and electrical properties of pulse electrodeposited Cu–graphene oxide composite coating for electrical contacts. *J. Mater. Sci.* **2017**, *52*, 1089–1105. [CrossRef]
11. Hidalgo-Manrique, P.; Lei, X.; Xu, R.; Zhou, M.; Kinloch, I.A.; Young, R.J. Copper/graphene composites: A review. *J. Mater. Sci.* **2019**, *54*, 12236. [CrossRef]
12. Sookhakian, M.; Ridwan, N.A.; Zalnezhad, E.; Yoon, G.H.; Azarang, M.; Mahmoudian, M.R.; Alias, Y. Layer-by-layer electrodeposited reduced graphene oxide-copper nanopolyhedra films as efficient platinum-free counter electrodes in high efficiency dye-sensitized solar cells. *J. Electrochem. Soc.* **2016**, *163*, D154–D159. [CrossRef]
13. Givalou, L.; Tsichlis, D.; Zhang, F.; Karagianni, C.-S.; Terrones, M.; Kordatos, K.; Falaras, P. Transition metal—Graphene oxide nanohybrid materials as counter electrodes for high efficiency quantum dot solar cells. *Catal. Today* **2019**. [CrossRef]
14. Kamboj, A.; Raghupathy, Y.; Rekha, M.Y.; Srivastava, C. Morphology, texture and corrosion behavior of nanocrystalline copper–graphene composite coatings. *JOM* **2017**, *69*, 1149–1154. [CrossRef]
15. Raghupathy, Y.; Kamboj, A.; Rekha, M.Y.; Narasimha Rao, N.P.; Srivastava, C. Copper–graphene oxide composite coatings for corrosion protection of mild steel in 3.5% NaCl. *Thin Solid Film* **2017**, *636*, 107–115. [CrossRef]
16. Li, S.; Song, G.; Fu, Q.; Pan, C. Preparation of Cu- graphene coating via electroless plating for high mechanical property and corrosive resistance. *J. Alloy. Compd.* **2019**, *777*, 877–885. [CrossRef]
17. Protich, Z.; Santhanam, K.S.V.; Jaikumar, A.; Kandlikar, S.G.; Wong, P. Electrochemical deposition of copper in graphene quantum dot bath: Pool boiling enhancement. *J. Electrochem. Soc.* **2016**, *163*, E166–E172. [CrossRef]

18. Jaykumar, A.; Santhanam, K.S.V.; Kandlikar, S.; Raya, I.B.P.; Raghupathi, P. Electrochemical deposition of copper on graphene with a high heat transfer coefficient. *ECS Trans.* **2015**, *66*, 55. [CrossRef]

19. Zhu, L.; Guo, X.; Liu, Y.; Chen, Z.; Zhang, W.; Yin, K.; Li, L.; Zhang, Y.; Wang, Z.; Sun, L. High-performance Cu nanoparticles/three-dimensional graphene/Ni foam hybrid for catalytic and sensing applications. *Nanotechnology* **2018**, *29*, 145703. [CrossRef]

20. Qin, L.; Xu, H.; Zhu, K.; Kang, S.-Z.; Li, G.; Li, X. Noble-metal-free copper nanoparticles/reduced graphene oxide composite: A new and highly efficient catalyst for transformation of 4-Nitrophenol. *Catal. Lett.* **2017**, *147*, 1315–1321. [CrossRef]

21. Gao, H.; Wang, Y.; Xiao, F.; Ching, C.B.; Duan, H. Growth of copper nanocubes on graphene paper as free-standing electrodes for direct hydrazine fuel cells. *J. Phys. Chem. C* **2012**, *116*, 7719–7725. [CrossRef]

22. Liu, C.; Zhang, H.; Tang, Y.; Luo, S. Controllable growth of graphene/Cu composite and its nanoarchitecture-dependent electrocatalytic activity to hydrazine oxidation. *J. Mater. Chem. A* **2014**, *2*, 4580–4587. [CrossRef]

23. Periasamy, A.P.; Liu, J.; Lin, H.-M.; Chang, H.-T. Synthesis of copper nanowire decorated reduced graphene oxide for electro-oxidation of methanol. *J. Mater. Chem. A* **2013**, *1*, 5973–5981. [CrossRef]

24. Muralikrishna, S.; Ravishankar, T.N.; Ramakrishnappa, T.; Nagaraju, D.H.; Krishna Pai, R. Non-noble metal graphene oxide-copper (II) ions hybrid electrodes for electrocatalytic hydrogen evolution reaction. *Environ. Prog. Sustain. Energy* **2016**, *35*, 565–573. [CrossRef]

25. He, T.; Zhang, C.; Du, A. Single-atom supported on graphene grain boundary as an efficient electrocatalyst for hydrogen evolution reaction. *Chem. Eng. Sci.* **2019**, *194*, 58–63. [CrossRef]

26. Yuan, J.; Yang, M.-P.; Zhi, W.-Y.; Wang, H.; Wang, H.; Lu, J.-X. Efficient electrochemical reduction of CO_2 to ethanol on Cu nanoparticles decorated on N-doped graphene oxide catalysts. *J. CO2 Util.* **2019**, *33*, 452–460. [CrossRef]

27. Shi, R.; Zhao, J.; Liu, S.; Sun, W.; Li, H.; Hao, P.; Li, Z.; Ren, J. Nitrogen-doped graphene supported copper catalysts for methanol oxidative carbonylation: Enhancement of catalytic activity and stability by nitrogen species. *Carbon* **2018**, *130*, 185–195. [CrossRef]

28. Sirijaraensre, J.; Khongpracha, P.; Limtrakul, J. Mechanistic insights into CO_2 cycloaddition to propylene oxide over a single copper atom incorporated graphene-based materials: A theoretical study. *Appl. Surf. Sci.* **2019**, *470*, 755–763. [CrossRef]

29. Sredojević, D.N.; Šljivančanin, Ž.; Brothers, E.N.; Belić, M.R. Formic Acid Synthesis by CO_2 Hydrogenation over Single-Atom Catalysts Based on Ru and Cu Embedded in Graphene. *Chem. Sel.* **2018**, *3*, 2631–2637. [CrossRef]

30. Back, S.; Lim, J.; Kim, N.-Y.; Kim, Y.-H.; Jung, Y. Single-atom catalysts for CO_2 electroreduction with significant activity and selectivity improvements. *Chem. Sci.* **2017**, *8*, 1090–1096. [CrossRef]

31. Öztürk Doğan, H.; Kurt Urhan, B.; Çepni, E.; Eryiğit, M. Simultaneous electrochemical detection of ascorbic acid and dopamine on Cu_2O/CuO/electrochemically reduced graphene oxide (CuxO/ERGO)-nanocomposite-modified electrode. *Microchem. J.* **2019**, *150*, 104157. [CrossRef]

32. Fu, J.; An, X.; Yao, Y.; Guo, Y.; Sun, X. Electrochemical aptasensor based on one step co-electrodeposition of aptamer and GO-CuNPs nanocomposite for organophosphorus pesticide detection. *Sens. Actuators B* **2019**, *287*, 503–509. [CrossRef]

33. Li, D.; Wang, C.; Zhang, H.; Sun, Y.; Duan, Q.; Ji, J.; Zhang, W.; Sang, S. A highly effective copper nanoparticle coupled with RGO for electrochemical detection of heavy metal ions. *Int. J. Electrochem. Sci.* **2017**, *12*, 10933–10945. [CrossRef]

34. Liu, C.; Qiao, C. Preparation of graphene-copper nanocomposite for electrochemical determination of cadmium ions in water. *Int. J. Electrochem. Sci.* **2017**, *12*, 8357–8367. [CrossRef]

35. Cui, D.; Su, L.; Li, H.; Li, M.; Li, C.; Xu, S.; Qian, L.; Yang, B. Non-enzymatic glucose sensor based on micro-/nanostructured Cu/Ni deposited on graphene sheets. *J. Electroanal. Chem.* **2019**, *838*, 154–162. [CrossRef]

36. Jiang, J.; Zhang, P.; Liu, Y.; Luo, H. A novel non-enzymatic glucose sensor based on a Cu-nanoparticle-modified graphene edge nanoelectrode. *Anal. Methods* **2017**, *9*, 2205–2210. [CrossRef]

37. Luo, J.; Jiang, S.; Zhang, H.; Jiang, J.; Liu, X. A novel non-enzymatic glucose sensor based on Cu nanoparticle modified graphene sheets electrode. *Anal. Chim. Acta* **2012**, *709*, 47–53. [CrossRef]

38. Balasubramanian, P.; Velmurugan, M.; Chen, S.-M.; Hwa, K.-Y. Optimized electrochemical synthesis of copper nanoparticles decorated reduced graphene oxide: Application for enzymeless determination of glucose in human blood. *J. Electroanal. Chem.* **2017**, *807*, 128–136. [CrossRef]

39. Shabnam, L.; Faisal, S.N.; Roy, A.K.; Haque, E.; Minett, A.I.; Gomes, V.G. Doped graphene/Cu nanocomposite: A high sensitivity non-enzymatic glucose sensor for food. *Food Chem.* **2017**, *221*, 751–759. [CrossRef]

40. Yang, Y.; Ma, N.; Bian, Z. Cu-Au/rGO nanoparticle based electrochemical sensor for 4-chlorophenol detection. *Int. J. Electrochem. Sci.* **2019**, *14*, 4095–4113. [CrossRef]

41. Dorraji, P.S.; Jalali, F. A nanocomposite of poly(melamine) and electrochemically reduced graphene oxide decorated with Cu nanoparticles: Application to simultaneous determination of hydroquinone and catechol. *J. Electrochem. Soc.* **2015**, *162*, B237–B244. [CrossRef]

42. Wang, H.; Wang, C.; Yang, B.; Zhai, C.; Bin, D.; Zhang, K.; Yang, P.; Du, Y. A facile fabrication of copper particle-decorated novel graphene flower composites for enhanced detecting of nitrite. *Analyst* **2015**, *140*, 1291–1297. [CrossRef] [PubMed]

43. Majidi, M.R.; Ghaderi, S. Hydrogen bubble dynamic template fabrication of nanoporous Cu film supported by graphene nanosheets: A highly sensitive sensor for detection of nitrite. *Talanta* **2017**, *175*, 21–29. [CrossRef] [PubMed]

44. Su, Z.; Tan, L.; Yang, R.; Zhang, Y.; Tao, J.; Zhang, N.; Wen, F. Cu-modified carbon spheres/reduced graphene oxide as a high sensitivity of gas sensor for NO2 detection at room temperature. *Chem. Phys. Lett.* **2018**, *695*, 153–157. [CrossRef]

45. Muralikrishna, S.; Cheunkar, S.; Lertanantawong, B.; Ramakrishnappa, T.; Nagaraju, D.H.; Surareungchai, W.; Balakrishna, R.G.; Reddy, K.R. Graphene oxide-Cu(II) composite electrode for non-enzymatic determination of hydrogen peroxide. *J. Electroanal. Chem.* **2016**, *776*, 9–65. [CrossRef]

46. Zhou, Q.; Su, X.; Ju, W.; Yong, Y.; Li, X.; Fu, Z.; Wang, C. Adsorption of H2S on graphane decorated with Fe, Co and Cu: A DFT study. *RSC Adv.* **2017**, *7*, 31457–31465. [CrossRef]

47. Mohammadi-Manesh, E.; Vaezzadeh, M.; Saeidi, M. Cu- and CuO-decorated graphene as a nanosensor for H2S detection at room temperature. *Surf. Sci.* **2015**, *636*, 36–41. [CrossRef]

48. Zheng, J.H.; Niu, S.F.; Lian, J.S. Carbon monoxide adsorption on cooper doped graphene systems: A DFT study. *Optoelectron. Adv. Mater. Rapid Commun.* **2014**, *8*, 1044–1049.

49. Li, X.; Chen, X.; Qi, J. First-principle theory calculations of CO_2 adsorption and activation by metal-graphene composite. *Harbin Gongye Daxue Xuebao J. Harbin Inst. Technol.* **2014**, *46*, 58–64.

50. Liu, Z.; Cheng, X.; Yang, Y.; Jia, H.; Bai, B.; Zhao, L. DFT study of N_2O adsorption onto the surface of M -decorated graphene oxide (M. = Mg, Cu or Ag). *Materials* **2019**, *12*, 2611. [CrossRef]

51. Choudhary, A.; Malakkal, L.; Siripurapu, R.K.; Szpunar, B.; Szpunar, J. First principles calculations of hydrogen storage on Cu and Pd-decorated graphene. *Int. J. Hydrog. Energy* **2016**, *41*, 17652–17656. [CrossRef]

52. Wong, J.; Yadav, S.; Tam, J.; Veer Singh, C. A van der Waals density functional theory comparison of metal decorated graphene systems for hydrogen adsorption. *J. Appl. Phys.* **2014**, *115*, 224301. [CrossRef]

53. Sigal, A.; Rojas, M.I.; Leiva, E.P.M. Is hydrogen storage possible in metal-doped graphite 2D systems in conditions found on earth? *Phys. Rev. Lett.* **2011**, *107*, 158701. [CrossRef] [PubMed]

54. Malček, M.; Cordeiro, M.N.D.S. A DFT and QTAIM study of the adsorption of organic molecules over the copper-doped coronene and circumcoronene. *Phys. E* **2018**, *95*, 59–70. [CrossRef]

55. Malček, M.; Bučinský, L.; Teixeira, F.; Cordeiro, M.N.D.S. Detection of simple inorganic and organic molecules over Cu-decorated circumcoronene: A combined DFT and QTAIM study. *Phys. Chem. Chem. Phys.* **2018**, *20*, 16021–16032. [CrossRef]

56. Düzenli, D.A. Comparative density functional study of hydrogen peroxide adsorption and activation on the graphene surface doped with N, B, S, Pd, Pt, Au, Ag, and Cu Atoms. *J. Phys. Chem. C* **2016**, *120*, 20149–20157. [CrossRef]

57. National Research Council (US) Committee on Copper in Drinking Water. *Copper in Drinking Water*; Health Effects of Excess Copper; National Academies Press: Washington, DC, USA, 2000; p. 5.

58. Brewer, G.J. Alzheimer's disease causation by copper toxicity and treatment with zinc. *Front. Aging Neurosci.* **2014**, *16*, 92. [CrossRef]

59. Schedin, F.; Geim, A.K.; Morozov, S.V.; Hill, E.W.; Blake, P.; Katsnelson, M.I.; Novoselov, K.S.; Morozov, S.; Novoselov, K. Detection of individual gas molecules adsorbed on graphene. *Nat. Mater.* **2007**, *6*, 652–655. [CrossRef]

60. Yang, T.; Zhao, X.; He, Y.; Zhu, H. *Graphene-Based Sensors*; Elsevier BV: Amsterdam, The Netherlands, 2018; pp. 157–174.

61. Molina, J.; Cases, F.; Moretto, L.M. Graphene-based materials for the electrochemical determination of hazardous ions. *Anal. Chim. Acta* **2016**, *946*, 9–39. [CrossRef]

62. Petsawi, P.; Yaiwong, P.; Laocharoensuk, R.; Ounnunkad, K. Determination of copper (II) and cadmium (II) in rice samples by anodic stripping square wave voltammetry using reduced graphene oxide/polypyrrole composite modified screen-printed carbon electrode. *Chiang Mai J. Sci.* **2019**, *46*, 322–336.

63. Tian, X.; Tan, Z.; Zhang, Z.; Zhan, T.; Liu, X. An electrochemical sensor based on an ionic liquid covalently functionalized graphene oxide for simultaneous determination of copper (II) and antimony (III). *Chem. Sel.* **2018**, *3*, 8252–8258. [CrossRef]

64. Ahour, F.; Taheri, M. Anodic stripping voltammetric determination of copper (II) ions at a graphene quantum dot-modified pencil graphite electrode. *J. Iran. Chem. Soc.* **2018**, *15*, 343–350. [CrossRef]

65. Wang, Y.; Zhao, S.; Li, M.; Li, W.; Zhao, Y.; Qi, J.; Cui, X. Graphene quantum dots decorated graphene as an enhanced sensing platform for sensitive and selective detection of copper (II). *J. Electroanal. Chem.* **2017**, *797*, 113–120. [CrossRef]

66. Sun, H.; Jia, Y.; Dong, H.; Fan, L. Graphene oxide nanosheets coupled with paper microfluidics for enhanced on-site airborne trace metal detection. *Microsyst. Nanoeng.* **2019**, *5*, 4. [CrossRef]

67. Zhang, Y.; Li, K.; Ren, S.; Dang, Y.; Liu, G.; Zhang, R.; Zhang, K.; Long, X.; Jia, K. Coal-derived graphene quantum dots produced by ultrasonic physical tailoring and their capacity for Cu(II) detection. *ACS Sustain. Chem. Eng.* **2019**, *7*, 9793–9799. [CrossRef]

68. Akhila, A.K.; Renuka, N.K. Coumarin-graphene turn-on fluorescent probe for femtomolar level detection of copper(ii). *N. J. Chem.* **2019**, *43*, 1001–1008. [CrossRef]

69. Nazerdeylami, S.; Ghasemi, J.B.; Badiei, A. Anthracene modified graphene oxide-silica as an optical sensor for selective detection of Cu2+ and I− ions. *Int. J. Environ. Anal. Chem.* **2019**, 1–16. [CrossRef]

70. Basiri, S.; Mehdinia, A.; Jabbari, A. Green synthesis of reduced graphene oxide-Ag nanoparticles as a dual-responsive colorimetric platform for detection of dopamine and Cu2+. *Sens. Actuators B* **2018**, *262*, 499–507. [CrossRef]

71. Wang, C.; Yang, F.; Tang, Y.; Yang, W.; Zhong, H.; Yu, C.; Li, R.; Zhou, H.; Li, Y.; Mao, L. Graphene quantum dots nanosensor derived from 3D nanomesh graphene frameworks and its application for fluorescent sensing of Cu 2+ in rat brain. *Sens. Actuators B* **2018**, *258*, 672–681. [CrossRef]

72. Song, F.; Ai, Y.; Zhong, W.; Wang, J. Detection of copper ions and glutathione based on off-on fluorescent graphene quantum dots. *J. China. Pharm. Univ.* **2018**, *49*, 87–92. [CrossRef]

73. Li, M.; Liu, Z.; Wang, S.; Calatayud, D.G.; Zhu, W.-H.; James, T.D.; Wang, L.; Mao, B.; Xiao, H.-N. Fluorescence detection and removal of copper from water using a biobased and biodegradable 2D soft material. *Chem. Commun.* **2018**, *54*, 184–187. [CrossRef] [PubMed]

74. Zheng, W.; Li, H.; Chen, W.; Zhang, J.; Wang, N.; Guo, X.; Jiang, X. Rapid detection of copper in biological systems using click chemistry. *Small* **2018**, *14*, 1703857. [CrossRef] [PubMed]

75. Shtepliuk, I.; Khranovskyy, V.; Yakimova, R. Combining graphene with silicon carbide: Synthesis and properties—A review. *Semicond. Sci. Technol.* **2016**, *31*, 113004. [CrossRef]

76. Shtepliuk, I.; Iakimov, T.; Khranovskyy, V.; Eriksson, J.; Giannazzo, F.; Yakimova, R. Role of the Potential Barrier in the Electrical Performance of the Graphene/SiC Interface. *Crystals* **2017**, *7*, 162. [CrossRef]

77. Yazdi, G.R.; Iakimov, T.; Yakimova, R. Epitaxial Graphene on SiC: A Review of Growth and Characterization. *Crystals* **2016**, *6*, 53. [CrossRef]

78. Shtepliuk, I.; Santangelo, M.F.; Vagin, M.; Ivanov, I.G.; Khranovskyy, V.; Iakimov, T.; Eriksson, J.; Yakimova, R. Understanding Graphene response to neutral and charged lead species: Theory and experiment. *Materials* **2018**, *11*, 2059. [CrossRef]

79. Santangelo, M.F.; Shtepliuk, I.; Filippini, D.; Ivanov, I.G.; Yakimova, R.; Eriksson, J. Real-time sensing of lead with epitaxial graphene-integrated microfluidic devices. *Sens. Actuators B* **2019**, *288*, 425–431. [CrossRef]

80. Shtepliuk, I.; Vagin, M.; Yakimova, R. Insights into the Electrochemical Behavior of Mercury on Graphene/SiC Electrodes. *C J. Carbon Res.* **2019**, *5*, 51. [CrossRef]

81. Yakimova, R.; Iakimov, T.; Syväjärvi, M. Process for Growth of Graphene. U.S. Patent US9150417B2, 6 October 2015.

82. Ivanov, I.G.; Hassan, J.U.; Iakimov, T.; Zakharov, A.A.; Yakimova, R.; Janzén, E. Layer-number determination in graphene on SiC by reflectance mapping. *Carbon* **2014**, *77*, 492–500. [CrossRef]

83. Vagin, M.Y.; Sekretaryova, A.N.; Ivanov, I.G.; Håkansson, A.; Iakimov, T.; Syväjärvi, M.; Yakimova, R.; Lundström, I.; Eriksson, M. Monitoring of epitaxial graphene anodization. *Electrochim. Acta* **2017**, *238*, 91–98. [CrossRef]

84. Yamaguchi, T.; Nomura, M.; Wakita, H.; Ohtaki, H. An extended x-ray absorption fine structure study of aqueous rare earth perchlorate solutions in liquid and glassy states. *J. Chem. Phys.* **1988**, *89*, 5153–5159. [CrossRef]

85. Sémon, L.; Boehme, C.; Billard, I.; Hennig, C.; Lützenkirchen, K.; Reich, T.; Rossini, I.; Wipff, G.; Roßberg, A.; Roßberg, A. Do Perchlorate and Triflate Anions Bind to the Uranyl Cation in an Acidic Aqueous Medium? A Combined EXAFS and Quantum Mechanical Investigation. *Chem. Phys. Chem.* **2001**, *2*, 591–598. [CrossRef]

86. Binnemans, K. Applications of tetravalent cerium compounds. In *Handbook on the Physics and Chemistry of Rare Earths*, 1st ed.; Gschneidner, K.A., Jr., Bünzli, J.-C.G., Pecharsky, V.K., Eds.; Elsevier: Amsterdam, The Netherlands, 2006; Volume 3, pp. 306–307. ISBN 9780080466729.

87. Scharifker, B.; Hills, G. Theoretical and experimental studies of multiple nucleation. *Electrochim. Acta* **1983**, *28*, 879–889. [CrossRef]

88. Frisch, M.J. Gaussian 16. In *Revision, B. 01*; Gaussian Inc.: Wallingford, CT, USA, 2016.

89. Adamo, C.; Barone, V. Toward reliable density functional methods without adjustable parameters: The PBE0 model. *J. Chem. Phys.* **1999**, *110*, 6158–6170. [CrossRef]

90. Perdew, J.P.; Burke, K.; Ernzerhof, M. Generalized Gradient Approximation Made Simple. *Phys. Rev. Lett.* **1996**, *77*, 3865–3868. [CrossRef]

91. Martin, G.; Sundermann, A. Correlation consistent valence basis sets for use with the Stuttgart–Dresden–Bonn relativistic effective core potentials: The atoms Ga–Kr and In–Xe. *J. Chem. Phys.* **2001**, *114*, 3408–3420. [CrossRef]

92. Lu, T.; Chen, F. Multiwfn: A multifunctional wavefunction analyzer. *J. Comput. Chem.* **2012**, *33*, 580–592. [CrossRef]

93. Santangelo, M.F.; Shtepliuk, I.; Filippini, D.; Puglisi, D.; Vagin, M.; Yakimova, R.; Eriksson, J. Epitaxial Graphene sensors combined with 3D-printed microfluidic chip for heavy metals detection. *Sensors* **2019**, *19*, 2393. [CrossRef]

94. Abbott, A.P.; El Ttaib, K.; Frisch, G.; McKenzie, K.J.; Ryder, K.S. Electrodeposition of copper composites from deep eutectic solvents based on choline chloride. *Phys. Chem. Chem. Phys.* **2009**, *11*, 4269–4277. [CrossRef]

95. Sebastian, P.; Valles, E.; Gomez, E. Copper electrodeposition in a deep eutectic solvent. First stages analysis considering Cu(I) stabilization in chloride media. *Electrochim. Acta* **2014**, *123*, 285–295. [CrossRef]

96. Jovic, V.D.; Jovic, G.M. Copper electrodeposition from a copper acid baths in the presence of PEG and NaCl. *J. Serb. Chem. Soc.* **2001**, *66*, 935–952. [CrossRef]

97. Che, C.Y.; Vagin, M.; Wijeratne, K.; Zhao, D.; Warczak, M.; Jonsson, M.P.; Crispin, X. Conducting polymer electrocatalysts for proton-coupled electron transfer reactions: Toward organic fuel cells with forest fuels. *Adv. Sustain. Syst.* **2018**, *2*, 1800021. [CrossRef]

98. Persson, I.; Persson, P.; Sandström, M.; Ullström, A.-S. Structure of Jahn–Teller distorted solvate copper (II) ions in solution, and in solids with apparently regular octahedral coordination geometry. *Chem. Soc. Dalton Trans.* **2002**, *7*, 1256–1265. [CrossRef]

99. Li, J.; Fisher, C.L.; Chen, J.L.; Bashford, D.; Noodleman, L. Calculation of redox potentials and pKa values of hydrated transition metal cations by a combined density functional and continuum dielectric theory. *Inorg. Chem.* **1996**, *35*, 4694. [CrossRef]

100. Smith, E.L.; Barron, J.C.; Abbott, A.P.; Ryder, K.S. Time resolved in situ liquidatomic force microscopy and simultaneous acoustic impedance electrochemical quartz crystal microbalance measurements: A study of Zn deposition. *Anal. Chem.* **2009**, *81*, 8466–8471. [CrossRef]

101. Shtepliuk, I.; Vagin, M.; Ivanov, I.G.; Iakimov, T.; Yazdi, G.R.; Yakimova, R. Lead (Pb) interfacing with epitaxial graphene. *Phys. Chem. Chem. Phys.* **2018**, *20*, 17105–17116. [CrossRef]

102. Ziegler, C.; Wielgosz, R.I.; Kolb, D.M. Pb deposition on n-Si (111) electrodes. *Electrochim. Acta* **1999**, *45*, 827–833. [CrossRef]
103. Tułodziecki, M.; Tarascona, J.-M.; Tabernab, P.L.; Guérya, C. Importance of the double layer structure in the electrochemicaldeposition of Co from soluble Co2+- based precursors in Ionic Liquid media. *Electrochim. Acta* **2014**, *134*, 55–66. [CrossRef]

Review

Atomic Layer Deposition of High-k Insulators on Epitaxial Graphene: A Review

Filippo Giannazzo [1,*], **Emanuela Schilirò** [1,*], **Raffaella Lo Nigro** [1], **Fabrizio Roccaforte** [1] and **Rositsa Yakimova** [2]

[1]   CNR-IMM, Strada VIII, 5 95121 Catania, Italy; raffaella.lonigro@imm.cnr.it (R.L.N.);
     fabrizio.roccaforte@imm.cnr.it (F.R.)
[2]   Department of Physics Chemistry and Biology, Linköping University, SE-58183 Linköping, Sweden;
     rositsa.yakimova@liu.se
*   Correspondence: filippo.giannazzo@imm.cnr.it (F.G.); emanuela.schiliro@imm.cnr.it (E.S.)

Received: 7 March 2020; Accepted: 27 March 2020; Published: 3 April 2020

Featured Application: Graphene-based electronics and sensing.

Abstract: Due to its excellent physical properties and availability directly on a semiconductor substrate, epitaxial graphene (EG) grown on the (0001) face of hexagonal silicon carbide is a material of choice for advanced applications in electronics, metrology and sensing. The deposition of ultrathin high-k insulators on its surface is a key requirement for the fabrication of EG-based devices, and, in this context, atomic layer deposition (ALD) is the most suitable candidate to achieve uniform coating with nanometric thickness control. This paper presents an overview of the research on ALD of high-k insulators on EG, with a special emphasis on the role played by the peculiar electrical/structural properties of the EG/SiC (0001) interface in the nucleation step of the ALD process. The direct deposition of Al_2O_3 thin films on the pristine EG surface will be first discussed, demonstrating the critical role of monolayer EG uniformity to achieve a homogeneous Al_2O_3 coverage. Furthermore, the ALD of several high-k materials on EG coated with different seeding layers (oxidized metal films, directly deposited metal-oxides and self-assembled organic monolayers) or subjected to various prefunctionalization treatments (e.g., ozone or fluorine treatments) will be presented. The impact of the pretreatments and of thermal ALD growth on the defectivity and electrical properties (doping and carrier mobility) of the underlying EG will be discussed.

Keywords: epitaxial graphene; atomic layer deposition; high-k insulators

1. Introduction

Graphene, the two-dimensional (2D) sp^2 allotropic form of carbon, has been the object of continuously increasing scientific and technological interest, starting from its first isolation in 2004 [1]. Among the different types of graphene materials considered so far, the epitaxial graphene (EG) grown by controlled high temperature graphitization of the (0001) face of hexagonal silicon carbide (6H- or 4H-SiC) [2–4] is a material of choice for advanced applications in electronics [5,6], high precision metrology [7] and environmental sensing [8]. The main advantage of this growth method over other commonly used approaches (such as chemical vapor deposition, CVD, on catalytic metals [9]) is the availability of high quality graphene directly on a semiconducting or semi-insulating substrate, i.e., ready for electronic devices fabrication, without the need of transfer processes [10,11] typically responsible for contaminations and damages [12,13].

Deposition of uniform ultra-thin insulators (especially high-k dielectrics) on graphene is a key step for the fabrication of graphene-based electronic devices [6,14–17]. In this context, the atomic layer deposition (ALD), owing to its sequential layer-by-layer growth mechanism [18], is the most suitable candidate to achieve a conformal growth with subnanometric control on the thickness. However, the lack of out-of-plane bonds or surface groups in the sp^2 lattice of graphene typically represents the main obstacle to films nucleation in the early stage of the ALD growth. This typically results in an inhomogeneous coverage of graphene surface, although the quality of the deposited films strongly depends on the graphene synthesis method, the graphene substrate and eventual transfer processes of graphene from the native substrate to foreign ones. As an example, for defect-free graphene flakes exfoliated from graphite, ALD growth was found to occur preferentially at the edges of the flakes [19], whereas, in the case of polycrystalline graphene grown by CVD on metals and transferred to insulating substrates, the ALD nucleation typically occurs at the grain boundaries and on wrinkles [20]. Interestingly, in the specific case of EG grown by thermal decomposition of SiC (0001), uniform ALD coverage has been observed on the monolayer (1 L) EG areas, whereas inhomogeneous growth has been found on the bilayer (2 L) of few layer regions [21]. These ALD nucleation issues common to all the graphene materials are typically circumvent by adopting surface preparation protocols, consisting of chemical prefunctionalization of graphene surface to introduce reactive (sp^3) sites, or by the predeposition of seed layers [22–32]. However, proper optimization of these processes is necessary in order to avoid structural damages and a degradation of graphene electronic properties.

Although the research on ALD for graphene devices integration has been recently the object of comprehensive review articles [33], to the best of our knowledge, a focused paper reviewing ALD of high-k insulators on the EG/SiC(0001) system is currently missing. This article aims to provide an overview on this topic, highlighting the specific approaches adopted to achieve uniform ALD growth in the case of EG. Section 2 provides an introduction to the peculiar structural and electrical properties of EG and its interface with the (0001) SiC substrate. Section 3 presents recent results on the direct deposition of Al$_2$O$_3$ thin films on the pristine EG, elucidating the key role played by the sp^3 hybridized buffer layer to obtain a homogeneous Al$_2$O$_3$ coverage on monolayer EG. Furthermore, the ALD on EG coated with different seeding layers, including oxidized metal films [34], high-k metal oxide thin films [35], self-assembled organic monolayers [36] or spin coated graphene oxide [37], will be presented in Section 4. Different prefunctionalization treatments of EG surface by exposure to reactive gases such as ozone [38] or XeF$_2$ [39] will be described in the Section 5. The impact of these surface preparation processes and of thermal ALD growth on the defectivity and electrical properties of the underlying graphene will be discussed. Finally, open research issues and perspectives in the field of ALD on graphene are presented in the Section 6.

2. Morphology and Interface Structure of Epitaxial Graphene on SiC (0001)

Graphene growth by thermal decomposition of hexagonal SiC relies on the interplay of three mechanisms: (i) the preferential Si sublimation from the topmost SiC layers, leaving an excess of carbon, (ii) the diffusion of these C atoms on the SiC surface and (iii) their reorganization in the 2D hexagonal graphene lattice. These mechanisms depend both on the annealing conditions (sample temperature and gas partial pressures in the chamber) and on the SiC surface termination. To date most of the studies on graphene growth have been carried out on the Si terminated (0001) face [40–43] and on the C terminated (000-1) face [40–43], due to the availability of large area SiC substrates with these orientations. Some experiments on the non-polar faces (1-100) and (11-20) have been recently reported as well [44]. Graphene films with very different structural and electronic properties (defectivity, thickness homogeneity, doping and mobility) have been obtained on these different crystal orientations, as a result of the different surface reconstructions during thermal decomposition. Multilayers of graphene rotationally disordered to each other and with respect to the substrate are typically obtained on the (000-1) face. On the contrary, monolayer or Bernal stacked few layers of graphene showing single crystalline epitaxial alignment with the SiC substrate are normally achieved on the (0001) face. This is due to the specific growth mechanism, mediated by the formation of an interfacial carbon layer, the so-called buffer layer (BL), with partial sp^3 hybridization with the Si face [45]. Figure 1 schematically depicts the structure of the BL, which is covalently bonded to Si atoms of the substrate with a large density of dangling bonds, and to the topmost monolayer of graphene with Van der Walls forces. This peculiar interface structure makes EG compressively strained, and the electrostatic interaction with the dangling bonds at the BL/SiC interface is responsible for a high n-type doping (approx. 10^{13} cm^{-2}) of the overlying EG [46,47].

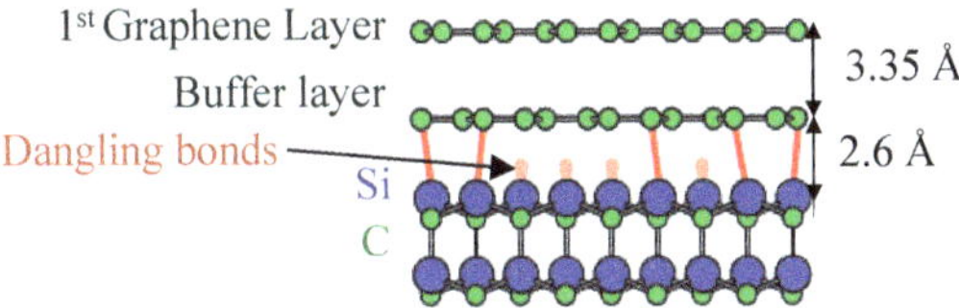

Figure 1. Illustration of the structure of monolayer EG grown on SiC (0001), showing the buffer layer partially bond to the Si face, with unsaturated dangling bonds at the interface.

Although thermal decomposition of the (0001) face allows a good control of the number of layers, achieving uniform 1 L graphene coverage on the entire SiC surface remains one of the main challenges in this research field. Besides having an impact on the EG electronic properties, thickness inhomogeneities of EG have also a critical impact on the nucleation of insulating films grown by ALD, as discussed in the Section 3.

The EG thickness uniformity on the (0001) face depends on the growth conditions (temperature and pressure) and on the substrate morphology, in particular the miscut angle, with better uniformity achieved for low miscut angle (nominally "on-axis") SiC. Under the ultra-high-vacuum (UHV) condition, EG formation has been achieved at temperatures as low as 1280 °C [3], but the topography and thickness distribution of the EG film was typically very inhomogeneous (composed of submicrometer 0 L, 1 L and 2 L patches) as shown by the atomic force microscopy (AFM) and low energy electron microscopy (LEEM) maps in Figure 2a,b. On the other hand, growth in inert gas (Ar) ambient at atmospheric pressure (approx. 900 mbar) allows it to greatly reduce the Si sublimation rate, raising the EG formation temperature at values as high as T = 1650 °C [3]. As shown in Figure 2c,d, the EG grown on "nominally" on-axis SiC(0001) is commonly composed of 1 L domains on the planar (0001) SiC terraces, separated by long and narrow 2 L or 3 L EG stripes at SiC step edges. Such steps are inherent of SiC crystal and their spacing is related to the miscut angle [48]. The preferential formation of 2 L and 3 L EG at their edges is related to the enhanced Si-desorption from these locations due to the weaker bonding in the SiC matrix. Although the EG grown under these atmospheric pressure/high temperature conditions is more homogeneous than EG grown under UHV, the presence of nanometric steps in the substrate morphology and of 1 L/2 L lateral junctions localized at these steps have been shown to cause a reduction of local electrical conductivity of EG [49,50].

Further improvement in the EG thickness homogeneity have been achieved by performing thermal decomposition of nominally on-axis 4H-SiC (0001) at a temperature of 2000 °C in inert gas (Ar) at atmospheric pressure in a radio frequency (RF) heated sublimation growth reactor [4]. By using specific well-controlled growth conditions (temperature distribution in the growth cell, temperature ramping up, and base pressure) monolayer EG coverage on most of the SiC surface has been obtained [21]. Figure 3a reports a reflectance map collected on a large area (30 μm × 30 μm) of EG grown under these optimized conditions. Reflectance mapping is a straightforward method to evaluate the number of layers distribution on large area EG samples by comparing the graphene thickness dependent reflected power with that of a bare 4H-SiC substrate [51]. Here the small yellow patches, corresponding to 2 L EG regions, covered only 1.3% surface and were surrounded by 1 L EG background on the 98.7% the area. Figure 3b,c shows the AFM morphology and the corresponding phase map on a 30 μm × 30 μm sample area. The morphological image shows the typical stepped surface of 4H-SiC (0001), and the variable contrast in the phase image provides information on the variation in the number of EG layers at different positions. In particular, the small elongated patches with higher phase contrast in Figure 3c correspond to the 2 L regions in the reflectance maps in Figure 3a.

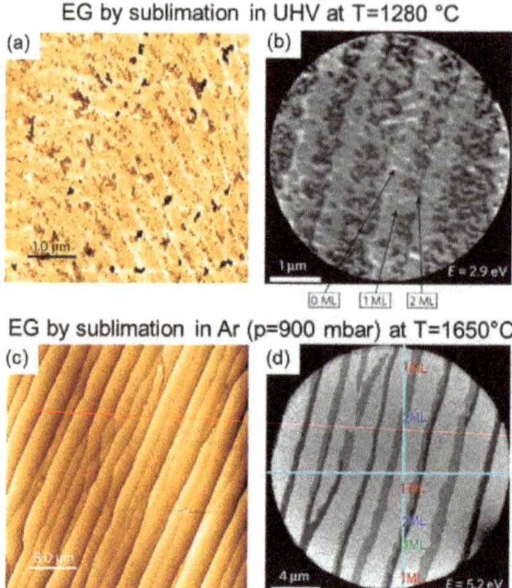

Figure 2. (**a**) Atomic force microscopy (AFM) morphology and (**b**) low energy electron microscopy (LEEM) map of EG grown on 6H–SiC(0001) by sublimation in ultra-high-vacuum (UHV) at a temperature T = 1280 °C. (**c**) AFM morphology and (**d**) LEEM map of EG on 6H–SiC(0001) obtained by sublimation in Ar ambient at atmospheric pressure (p = 900 mbar) and at a temperature T = 1650 °C. Darker LEEM contrast corresponds to a larger number of graphene layers. Images adapted with permission from Ref. [21], Copyright Nature Publishing Group 2009.

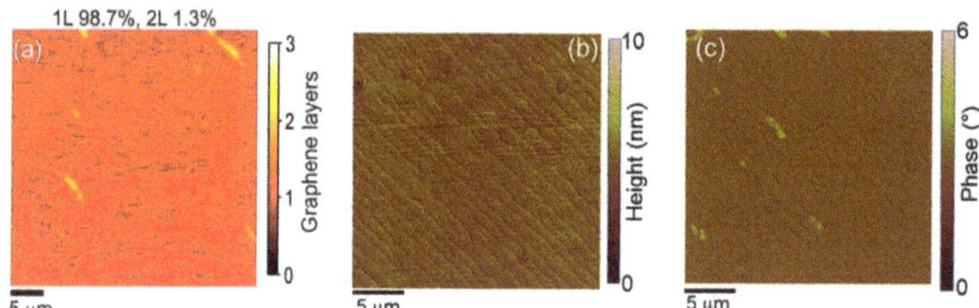

Figure 3. (**a**) Reflectance map of as-grown EG collected on a 30 µm × 30 µm sample area. The red contrast background is associated with 1 L graphene (98.7% of the total area) and the yellow elongated patches to 2 L graphene (1.3% of the total area). (**b**) AFM morphology and (**c**) phase contrast map on a 30 µm × 30 µm sample area. The small elongated patches with higher phase contrast correspond to 2 L EG. Images adapted with permission from Ref. [21], copyright Wiley 2019.

After introducing the peculiar structural and morphological properties of EG/SiC(0001), an overview of recent literature results in the ALD of high-k insulators on this material system would be provided in the next sections. Table 1 is a synoptic table including data from selected reference papers, including the specs of the EG material, the type of EG surface preparation adopted for the subsequent ALD process, the ALD growth details (precursors, temperature and number of cycles), the properties of the deposited dielectric material (thickness and uniformity), as well as the effect of the surface preparation/ALD process on the structural properties of EG.

Table 1. Overview of literature reports on atomic layer deposition (ALD) of dielectric materials on EG. Listed are the specs on EG growth process and number of graphene layers, the type of EG surface preparation used to achieve uniform ALD, the used ALD precursors and deposition temperatures, the thickness and uniformity of the deposited dielectric material, and whether EG was damaged by the surface preparation and/or the ALD process.

Ref	EG/SiC(0001) Specs	ALD Process				EG Damage
		EG Surface Prep	Precursors	T_{ALD} (°C)	Insulator	
[38]	EG growth at 1650 °C in Ar 1000 mbar, 1 L on SiC terraces, 2 L–3 L at SiC steps	None	Trimethylaluminum/H_2O, 500 cycles	200, 300, 350	Al_2O_3 (50 nm), not closed	No
[21]	EG growth at 2000 °C in Ar 900 mbar, 1 L EG (>98%) with submicron 2 L patches	None	Trimethylaluminum/H_2O, 500 cycles	250	Al_2O_3 (12 nm), closed on >98% surface	No
[34]	EG growth at 1600 °C in Ar 600 Torr, 1 L to 3 L EG	Oxidized Al seed layer, 2 nm	Triethylaluminum/H_2O	150, 300	Al_2O_3, closed	No
		Oxidized Ti seed layer, 2 nm	Titanium(IV) i-propoxide/H_2O	120, 250	TiO_2, closed	No
		Oxidized Ta seed layer, 2 nm	Pentakis(dimethylamino)tantalum(V)	150, 300	Ta_2O_5, closed, high roughness	Yes
[35]	EG growth at 1600 °C in Ar 900 mbar, Few layers EG	PVD Al_2O_3 seed layer, 2 nm	Triethylaluminum/H_2O	300	Al_2O_3 (8 nm) + Al_2O_3 seed (2 nm), closed	No
			Tetrakis(dimethylamino)hafnium/H_2O	250	HfO_2 (8 nm) + Al_2O_3 seed (2 nm), closed	No
		PVD HfO_2 seed layer, 2 nm	Triethylaluminum/H_2O	300	Al_2O_3 (8 nm) + HfO_2 seed (2 nm), closed	No
			Tetrakis (dimethylamino)hafnium/H_2O	250	HfO_2 (8 nm) + HfO_2 seed (2 nm), closed	No
		PVD SiO_2 seed layer, 2 nm	Tetrakis(dimethylamino)hafnium/H_2O	250	HfO_2 (10 nm) + SiO_2 seed (2 nm), closed	No
[36]	EG growth at 1350 °C in UHV, mixed 1 L–2 L EG	Organic seed layer: 1 L of perylene-3,4,9,10-tetracarboxylic dianhydride (PTCDA)	Trimethylaluminum/H_2O, 25 cycles	100	Al_2O_3 (2.3 nm), closed	No
			Tetrakis(diethylamido)hafnium(IV), 25 cycles	100	HfO_2 (2.5 nm), closed	No
[38]	EG growth at 1650 °C in Ar 1000 mbar, 1 L on SiC terraces, 2 L–3 L at SiC steps	O_3 prefunctionalization at T = 250 °C	Trimethylaluminum/H_2O, 500 cycles	250	Al_2O_3 (50 nm), closed	No
		O_3 prefunctionalization at T = 350 °C	Trimethylaluminum/H_2O, 500 cycles	350	Al_2O_3 (50 nm), closed	Yes
[39]	EG growth at 1650 °C in Ar, 100 mbar	Fluorine prefunctionalization by XeF_2 gas (from 0 to 200 s)	Trimethylaluminum/H_2O	225	Al_2O_3 (~15 nm), closed	No

3. Direct ALD on Pristine Epitaxial Graphene

Early studies on thermal ALD of high-k insulators (like Al_2O_3 or HfO_2) on pristine EG samples typically resulted in a non-uniform coverage, with poor or no oxide nucleation in the vicinity of the step edges [34,38]. As an example, Speck. et al. [38] investigated direct thermal ALD of Al_2O_3 on EG grown at 1650 °C in Ar (p = 900 mbar) [3]. Figure 4 illustrates the morphologies of the Al_2O_3 films obtained after 500 ALD cycles (corresponding to 50 nm Al_2O_3) at different deposition temperatures of 200 °C (a), 300 °C (b) and 350 °C (c), using trimethylaluminum (TMA) as the aluminum source, and water as the oxidant.

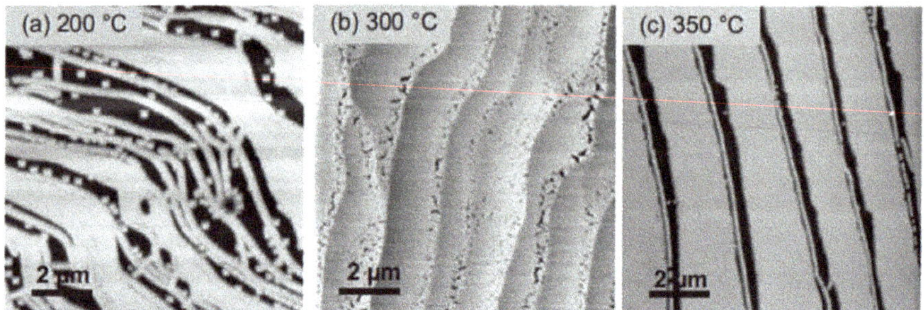

Figure 4. AFM images of Al_2O_3 grown directly on the pristine EG surface by 500 thermal ALD cycles of TMA/water (i.e., 50 nm Al_2O_3) at different temperatures: (**a**) 200 °C, (**b**) 300 °C and (**c**) 350 °C. Images adapted with permission from Ref. [38], copyright Wiley 2010.

Poor or no oxide nucleation was observed in the vicinity of the step edges. As discussed in the Section 2, samples with 1 L EG on the (0001) terraces typically exhibit 2 L or few layers EG stripes at the step edges between terraces [3]. Hence, the Al_2O_3 uncovered areas were associated by the authors to the presence of 2 L or few layer EG underneath. However, no clear explanation of the different ALD nucleation and growth behavior on 1 L and 2 L EG was provided. Similarly, Robinson et al. [34] reported that the direct thermal ALD of Ta_2O_5, and TiO_2 results in non-uniform coverage of EG up to a maximum deposition temperature (300 °C). More recently, Schilirò et al. [21] employed highly uniform 1 L EG samples (>98% monolayer coverage) as substrates for direct thermal ALD of Al_2O_3 at a temperature of 250 °C with TMA and H_2O precursors. Figure 5 shows the morphological and structural characterization of an Al_2O_3 film obtained after 190 deposition cycles, corresponding to a nominal film thickness of 15 nm, evaluated from the 0.08 nm/cycle deposition rate on a reference silicon substrate. Figure 5a depicts a representative morphological image by AFM on a 20 µm × 20 µm scan area. The Al_2O_3 film is conformal with the topography of the EG/4H-SiC surface, except for some small depressions showing the same elongated shape of the 2L graphene patches (see, for comparison, Figure 3). A high resolution cross-sectional TEM image of the as-deposited Al_2O_3 film is reported in Figure 5b, where the 1 L EG plus the underlying BL can be clearly identified, and the amorphous Al_2O_3 layer shows uniform contrast, indicating a uniform Al_2O_3 density. The measured Al_2O_3 thickness was 12 nm, which is thinner than the nominal one and was ascribed to a lower growth rate of Al_2O_3 on EG surface in the early stages of the deposition process.

Raman spectroscopy measurements were carried out both on the virgin EG sample and after the Al_2O_3 deposition, in order to evaluate the changes induced by the thermal ALD process at 250 °C on the structural quality and doping/strain of the underlying EG. Two representative Raman spectra for the two cases, respectively, are reported in Figure 5c. The characteristic G and 2D peaks of graphene exhibit single Lorentzian shape, and the FWHM of the 2D peaks in these representative spectra are consistent with the 1 L nature of EG [52]. The small changes in the positions of the G and 2D peaks after the Al_2O_3 deposition indicate that the ALD process did not significantly affect the doping and strain of the EG. The features in the 1200–1500 cm^{-1} range were related to the interfacial BL, whereas no significant increase of the the intensity of the disorder-related D peak (1300 cm^{-1}) was observed, indicating that no defects were introduced by the ALD process.

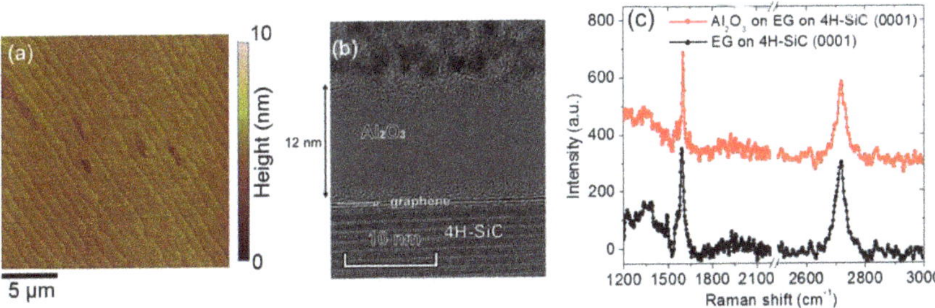

Figure 5. (a) AFM morphology (20 µm × 20 µm scan area) of Al$_2$O$_3$ directly grown on 1 L EG on SiC by 190 water/TMA ALD cycles. (b) Cross-sectional TEM image of the Al$_2$O$_3$ layer. (c) Representative Raman spectra of virgin EG and after the Al$_2$O$_3$ deposition. Images adapted with permission from Ref. [21], copyright Wiley 2019.

The electrical quality of the insulating layer was also evaluated by conductive atomic force microscopy (C-AFM) for current mapping and local current-voltage (I–V) analyses [21]. A morphology map of the scanned area is reported in Figure 6a, which includes both uniform Al$_2$O$_3$ on 1 L EG and Al$_2$O$_3$ on a 2 L EG patch. Figure 6b shows a current map collected on this area with a positive value of the tip bias V_{tip} = 6 V with respect to EG. While uniform low current values were detected through the 12 nm Al$_2$O$_3$ film onto 1 L EG, the presence of high current spots was observed in the 2 L EG region. Figure 6c illustrates two representative local current–voltage characteristics collected by the C-AFM probe on Al$_2$O$_3$ in the 1 L and 2 L EG regions. While current smoothly increased with the bias for Al$_2$O$_3$ on 1 L EG, an abrupt rise of current was observed for V_{tip} > 6 V in the case of Al$_2$O$_3$ on 2 L EG. This locally enhanced conduction in the 2 L EG area was justified by the less compact Al$_2$O$_3$ structure and the lower Al$_2$O$_3$ thickness detected in these regions. By adopting a simplified planar capacitor model for the tip/Al$_2$O$_3$/EG system, a breakdown field > 8 MV cm^{-1} was estimated for the 12 nm Al$_2$O$_3$ on 1 L EG.

The superior homogeneity of Al$_2$O$_3$ deposition on 1 L EG areas indicates a higher reactivity of 1 L EG with respect to 2 L or few layers EG, which was explained in terms of the higher n-type doping and strain caused by the interfacial BL on a single graphene overlayer with respect to 2 L [21]. Recent experimental and theoretical investigations demonstrated that the interaction of polar water molecules with graphene depends on the Fermi level of graphene, i.e., its doping [53]. In particular, ab-initio calculations of the adsorption energy (E$_a$) for water molecules (the co-reactant of the ALD growth) on 1 L graphene as a function of doping predicted an increase of E$_a$ from 127 meV for neutral graphene to 210 meV for highly n-type doped (10^{13} cm^{-2}) graphene [21]. The time of residence of a water molecule on graphene at a temperature T depends on E$_a$ as exp(E$_a$/k$_B$T), where k$_B$ is the Boltzmann constant. As a result, for the ALD process temperature (T = 250 °C), the residence time of physisorbed water molecules on the highly n-type doped EG was approximately six times higher than in the case of neutral graphene. The longer residence time provides, in turns, a larger number of reactive sites for Al$_2$O$_3$ formation during subsequent pulses of the Al precursor.

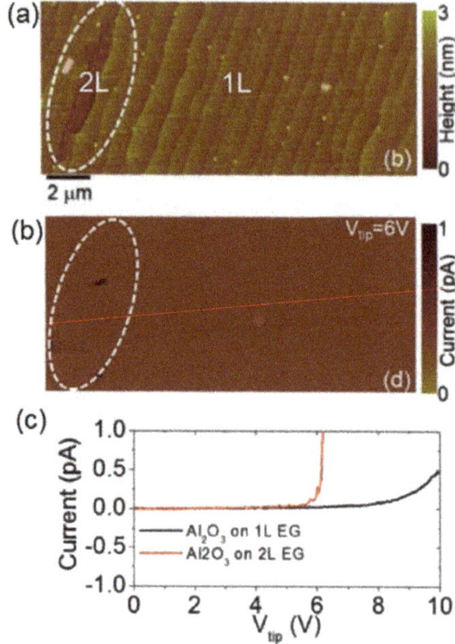

Figure 6. Conductive atomic force microscopy (C-AFM) local current mapping through an Al_2O_3 thin film deposited onto EG on axis 4H-SiC(0001). (**a**) Morphology of the probed sample area, including both uniform Al_2O_3 on 1 L EG and Al_2O_3 on a 2 L EG patch. (**b**) Current map collected on this area with a tip bias V_{tip} = 6 V. (**c**) Two representative local current-voltage characteristics collected by the C-AFM probe on Al_2O_3 in the 1 L and 2 L EG regions. Images adapted with permission from Ref. [21], copyright Wiley 2019.

4. ALD on EG With a Seeding-Layer

The above discussed literature results show that uniform and conformal high-k dielectrics could be obtained by direct ALD on monolayer EG areas. However, the availability of 100% 1 L EG coverage by thermal decomposition of SiC still represents a major challenge. For this reason, in many cases, the use of an intermediate seeding layer is the preferred solution to promote homogeneous ALD nucleation on the EG surface. In the following, the main seeding layer processes adopted so far will be presented, discussing the advantages and disadvantages in terms of their impact on the EG structural and electrical properties.

4.1. Oxidized Metal Seed Layer

Physical vapor deposition (PVD) of ultrathin metal films followed by oxidation is one of the most straightforward ways to create a seeding layer on the graphene surface [22]. Robinson et al. [34] investigated thermal ALD of different insulators (Al_2O_3, TiO_2 and Ta_2O_3) using oxidized metal films as seeding layers in the specific case of few layers EG/SiC(0001). Figure 7a,b shows the morphology of Al_2O_3 grown at 150 and 300 °C, respectively, with an oxidized Al (AlOx) seed layer. The Al_2O_3 film uniformity and coverage was found to be significantly improved by increasing the deposition temperature. Deposition of TiO_2 on EG with an oxidized Ti (TiOx) seed layer resulted in conformal and continuous films both at 120 °C (Figure 7c) and 250 °C (Figure 7d). Finally, ALD grown Ta_2O_5 films with an oxidized Ta (TaOx) seed layer showed a high roughness for deposition temperatures of 150 °C (Figure 7e) and 300 °C and (Figure 7f). The impact of the seeded thermal ALD on the EG structural properties (i.e., the defects density) was also qualified by Raman spectroscopy. Figure 7g shows typical Raman spectra of the D and G peaks for AlOx/Al_2O_3 deposited at 300 °C, TiOx/TiO_2

deposited at 250 °C, and TaOx/Ta$_2$O$_5$ deposited at 300 °C. These results show that ALD of high-k dielectric materials with an oxidized metal seed layer generally have little impact on the structural integrity of EG, with the highest EG defectivity obtained in the case of TaOx/Ta$_2$O$_5$, probably generated during oxidation of the Ta seed layer.

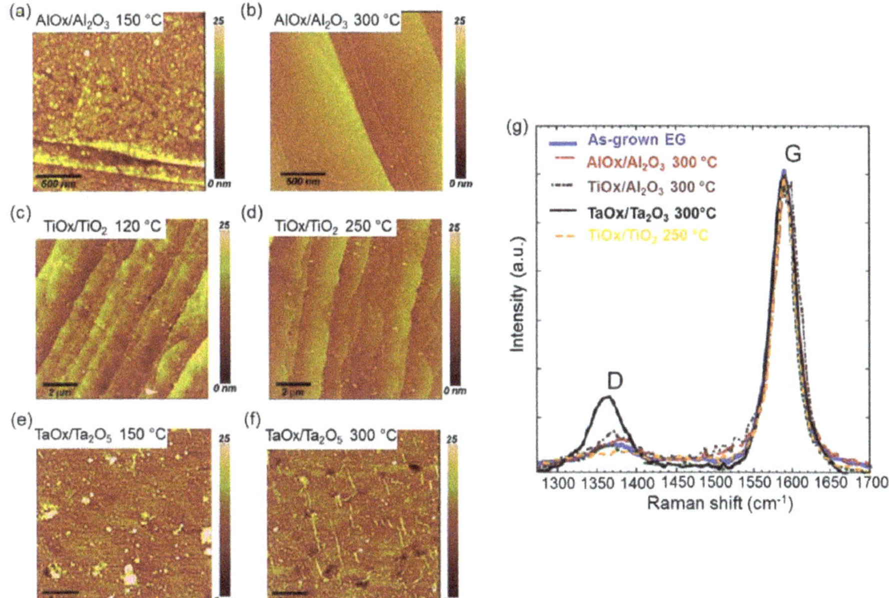

Figure 7. AFM morphology of Al$_2$O$_3$ thin films grown by ALD at 150 °C (**a**) and 300 °C (**b**) on EG/SiC with an oxidized Al seed layer. Morphology of TiO$_2$ grown by ALD at 120 °C (**c**) and 250 °C (**d**) on EG/SiC with an oxidized Ti seed layer. Morphology of Ta$_2$O$_5$ grown by ALD at 150 °C (**e**) and 300 °C (**f**) on EG/SiC with an oxidized Ta seed layer. (**g**) Raman spectra of as-grown EG and after ALD deposition of Al$_2$O$_3$, TiO$_2$ and Ta$_2$O$_5$ with different seed layers. Al$_2$O$_3$ and TiO$_2$ deposition introduces only a small amount of disorder in EG, whereas Ta$_2$O$_5$ deposition appears to significantly degrade the quality of the underlying EG. Images adapted with permission from Ref. [34], Copyright American Chemical Society 2010.

In spite of the limited increase in the EG defectivity, a reduction of the electron mobility has been reported in most of the cases after ALD of high-k dielectrics seeded by an oxidized metal [34]. This was ascribed to the poor structural quality and substoichiometric composition of the seed layer (typically due to an incomplete oxidation), leading to charge trapping phenomena and increased electron scattering by charged impurities.

4.2. Deposited High-k Metal-Oxide Seed Layers

As discussed above, the metal to metal-oxide phase transition occurring in the case of the oxidized metal seeding layers can be a source of mobility degradation in EG. To avoid these issues, the deposition of the seed-layer directly on EG from a high-purity oxide source has been also considered.

As an example, Hollander et al. [35] employed 2–3 nm seed layers of SiO$_2$, Al$_2$O$_3$ or HfO$_2$ deposited via nonreactive electron beam physical vapor deposition (EBPVD) at a pressure <10^{-6} Torr. Immediately following the seed-layer evaporation, 8–10 nm of Al$_2$O$_3$ or HfO$_2$ were deposited by ALD to complete the dielectric stack.

Figure 8a shows Raman spectra acquired on as-grown EG samples, after EBPVD oxide deposition and after the complete oxide-seeded ALD process. Minimal changes in the D/G ratio were observed among the different samples, whereas the blue-shift of the G peak after seed-layer and ALD was ascribed to an increase of EG doping.

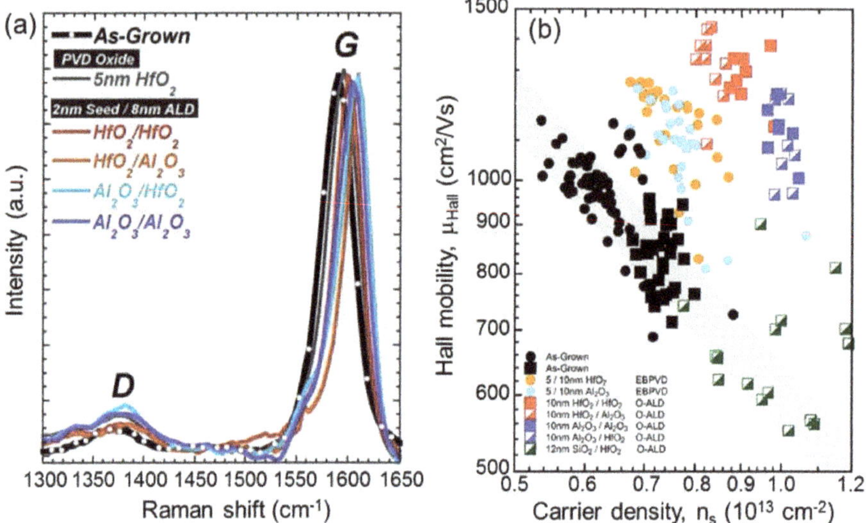

Figure 8. (a) Raman spectra of as-grown EG, after physical vapor deposition (PVD) of 5 nm HfO_2 and after HfO_2-seeded (or Al_2O_3-seeded) ALD of HfO_2 and Al_2O_3. (b) Hall mobility μ_{Hall} vs. carrier density n_s measured on as-grown EG, after electron beam physical vapor deposition (EBPVD) of oxide seed-layers and the complete oxide-seeded ALD (O-ALD) process. Images adapted with permission from Ref. [35], Copyright American Chemical Society 2010.

Hall effect measurements under high vacuum at 300 K were also carried out on EG prior to and following the metal-oxide seeded ALD in order to evaluate the effect of the deposited dielectrics on carrier transport properties, i.e., the carrier density (n_s) and Hall mobility (μ_{Hall}) [35].

Figure 8b shows the correlation between μ_{Hall} and n_s measured on as-grown EG, after EBPVD of oxides and the complete oxide-seeded ALD (O-ALD). As-grown EG samples presented μ_{Hall} values of 700–1100 $cm^2V^{-1}s^{-1}$ and n_s values of $(5–8) \times 10^{12}$ cm^{-2} (n-type), typical of Si-face EG. Figure 8 demonstrates that the high-k seeded dielectrics deposited by PVD and O-ALD resulted in an increase in n_s and an increase in μ_{Hall}. Conversely, HfO_2 seeded by a low-k insulator (SiO_2) resulted in a decrease of μ_{Hall} and the increase in n_s. The measured increase of μ_{Hall} with deposition of high-k seed was attributed to dielectric screening, i.e., a reduction in scattering by remote charged impurities [54]. In particular, HfO_2 seeded O-ALD dielectrics enhance carrier mobility by an estimated 57%–73%, while Al_2O_3 seeded O-ALD dielectrics increase mobility by 43%–52%.

4.3. Self-Assembled Organic Monolayer

Alaboson et al. [36] demonstrated an alternative ALD seeding layer based on organic monolayers of perylene-3,4,9,10-tetracarboxylic dianhydride (PTCDA). EG on SiC(0001) grown by sublimation in UHV conditions at 1350 °C was used in this study. PTCDA was deposited on the EG surface via gas-phase sublimation in ultrahigh vacuum (UHV), resulting in a highly uniform and ordered self-assembled monolayer. Two typical AFM images of the bare EG and PTCDA/EG surfaces are presented in Figure 9a,c, respectively, showing nearly identical roughness. Figure 9b,d shows two representative AFM images of Al_2O_3 deposited on the bare EG (b) and on PTCDA/EG (d) using

25 cycles of TMA and H_2O at a temperature of 100 °C. While a discontinuous Al_2O_3 film was observed on the bare EG, the use of the PTCDA seeding layer resulted in a uniform Al_2O_3 film, conformal with the underlying terraces of the EG surface. Figure 9e shows the evolution of the Al_2O_3 film thickness as a function of ALD growth cycle evaluated by spectroscopic ellipsometry on the bare EG, on PTCDA/EG and on a reference SiO_2 surface. The Al_2O_3 ALD on the bare EG surface was initially inhibited for 3–4 ALD cycles, indicating a lack of reactive sites for the ALD precursors, as compared to the case of the SiO_2 surface. On the other hand, nucleation on the PTCDA-seeded EG surface was more efficient with ALD growth showing a linear behavior after only 1–2 ALD cycles, similarly to the case of the SiO_2 surface with a high density of reactive sites.

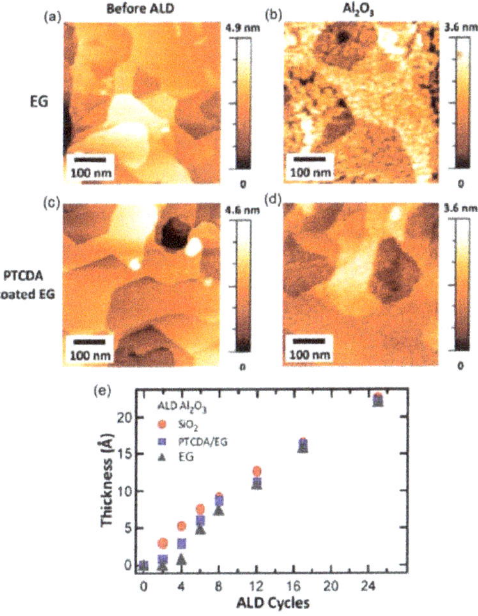

Figure 9. (a) Representative AFM image of an EG surface prepared by UHV graphitization. (b) AFM images of EG surface after ALD of 25 cycles of Al_2O_3. (c) AFM images of EG surface immediately after PTCDA deposition and (d) following ALD of 25 cycles of Al_2O_3. (e) Al_2O_3 film thickness as a function of ALD growth cycle evaluated by spectroscopic ellipsometry on the bare EG, on PTCDA/EG and on a reference SiO_2 surface. Images adapted with permission from Ref. [36], Copyright American Chemical Society 2011.

To probe the adhesion and the insulating properties of the deposited ALD films, C-AFM imaging was employed. The morphology and current maps in Figure 10a,b show that for a thin Al_2O_3 film (25 ALD cycles) on bare EG, conductive defects were produced after only one C-AFM scan in contact mode (contact force 20 nN, tip/sample voltage V = 0.3 V), resulting in localized conduction spots in the current map. This defect density was further increased after multiple C-AFM scans. Figure 10c,d shows the C-AFM morphology and current maps acquired under the same measurement conditions on the Al_2O_3 film grown using the same number of ALD cycles on PTCDA/EG. These maps demonstrate the excellent morphological and uniform insulating properties of Al_2O_3 on PTCDA seeded EG, which are maintained even after multiple C-AFM scans.

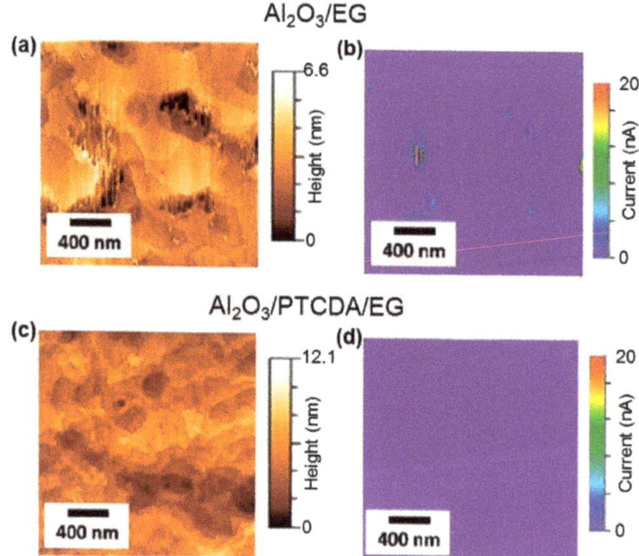

Figure 10. C-AFM morphology (**a**) and current map (**b**) of Al$_2$O$_3$ deposited by 25 ALD cycles on the bare EG surface. Morphology (**c**) and current map (**d**) of Al$_2$O$_3$ deposited with the same number of cycles on the PTCDA/EG surface. Images adapted with permission from Ref. [36], copyright American Chemical Society 2011.

5. ALD on Prefunctionalized EG

Different prefunctionalization processes of the EG surface by direct exposure to reactive gas species have been reported so far.

5.1. Ozone Prefunctionalization

Speck et al. [38] evaluated the effect of in-situ ozone (O$_3$) pretreatment on the uniformity of ALD grown Al$_2$O$_3$ onto EG. Firstly, a preliminary study of the influence of different O$_3$ dosing conditions (i.e., number of O$_3$ pulses and temperature) on the structural properties of EG was carried out. Figure 11a shows the comparison between two C1s core level X-ray photoelectron spectroscopy (XPS) measurements on as-grown EG and after exposure to 20 O$_3$ pulses at a temperature of 250 °C, respectively. The two spectra show two prominent peaks associated to the EG and to the SiC substrate (covered by EG), respectively, and a shoulder associated to the interfacial BL. The two spectra are almost perfectly overlapped, indicating a negligible effect of the O$_3$ treatment at 250 °C on EG structural and chemical properties. Figure 11b shows a C1s core level spectrum acquired on EG after 20 O$_3$ pulses at a higher temperature of 350 °C, with a deconvolution of the different spectral contributions, associated to the SiC substrate, to EG and to the BL. In addition to the components present in the pristine EG sample and in the O$_3$ -functionalized one at 250 °C (Figure 11a), another peak at lower binding energy, associated to SiC uncovered by EG, was observed in the EG sample subjected to the 20 O$_3$ pulses at 350 °C. From the XPS peaks intensity ratio, these bare SiC regions due to partial etching of EG by O$_3$, were found to correspond to 45% of EG surface. The degradation of the structural integrity of EG after this O$_3$ pretreatment (20 pulses at 350 °C) was also confirmed by Raman spectroscopy, showing a strong increase of the defects-related D peak and a large broadening of the G peak. Noteworthy, further experiments showed that EG damage can be mitigated by properly reducing the number of O$_3$ pulses [38]. As an example, the percentage of etched EG at 350 °C was found to be reduced to 2.5% by using only 2 O$_3$ pulses. Figure 11c shows a representative AFM image of a 50 nm thick Al$_2$O$_3$ film obtained by 500 TMA/water cycles on the EG sample prefunctionalized using 20 O_3 pulses at 250 °C, i.e., the non-destructive conditions illustrated in Figure 11a. Although a uniform Al_2O_3 coating was obtained on the entire EG surface, the higher roughness of the oxide film in the 2 L EG region close to the SiC steps indicated a limit for further reduction of O_3 exposure and temperature.

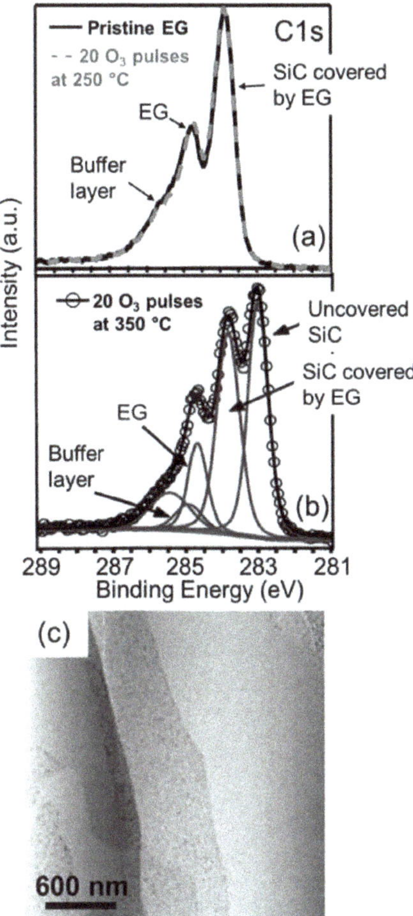

Figure 11. (**a**) XPS spectra from C1s core-level for pristine EG on SiC(0001) (solid line) and after 20 O_3 pulses at 250 °C. The two spectra are overlapped indicating negligible etching of graphene by the O_3 treatment. (**b**) C 1s core level spectrum of EG after 20 O_3 pulses at 350 °C, showing the appearance of an additional peak associated to uncovered SiC, which indicates partial etching of EG. (**c**) AFM image of a 50 nm thick Al_2O_3 film grown by 500 TMA/water cycles on EG pretreated with 20 O_3 pulses at 250 °C. Figures adapted with permission from Ref [38], copyright Wiley 2010.

5.2. Fluorine Prefunctionalization

Wheeler et al. [39] investigated fluorine prefunctionalization of EG for uniform deposition of thin high-k dielectrics. Due to its high electronegativity (4.0) and its ability to adhere to carbon surfaces [55], fluorine is a suitable reactive species to enhance graphene surface reactions with ALD precursors. Fluorination of the EG grown onto on-axis 6H–SiC (0001) was carried out at room temperature exposing the EG surface to XeF_2 gas pulses, with the total fluorine dosing time varied in the range from 0 to

200 s. Core level XPS was used to preliminary determine the bonding characteristics of fluorine with the EG surface with increasing XeF_2 exposure time. Figure 12a,d shows the C1s XPS spectra acquired on as-grown EG (a) and on EG subjected to 40 s (b), 120 s (c) and 200 s (d) XeF_2 exposure times. For as-grown EG (a), the contributions associated to the SiC substrate, to EG and to the interfacial buffer layer were visible. A very similar XPS spectrum was found after 40 s XeF_2 exposure (b), whereas the appearance of a feature at 288.5 eV, associated to C–F bonds, could be observed after 120 s (c) and 200 s (d) XeF_2 exposure times. Such bonds are created by breaking the sp^2 symmetry of the EG lattice resulting in carbon sp^3 bonds with fluorine. A decrease in the EG peak intensity was observed corresponding to the appearance of C-F bonds, which further supports the transition from sp^2 to sp^3 bonded carbon on the surface. For the largest exposure time (d), an additional small peak associated to $C–F_2$ bonds was detected at 290 eV. The formation of $C–F_2$ was accompanied by a significant deformation of the EG lattice, leading to a degradation of EG electrical properties. Hence, it was chosen to work in a fluorination regime where C–F bonding only occurred. The C-F sp^3 bond configuration was found to provide additional reaction sites of oxide nucleation and growth during the subsequent ALD process. After the initial assessment of the effect of XeF_2 exposure on EG chemical properties, Al_2O_3 films (15 nm thickness) were deposited by thermal ALD at 225 °C using TMA/H_2O precursors.

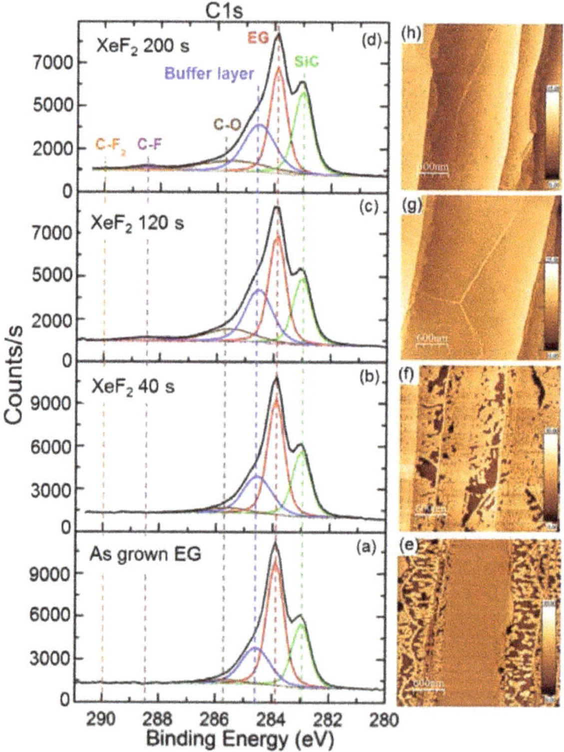

Figure 12. XPS C1s spectra on as-grown EG (**a**) and after XeF_2 exposure for 40 s (**b**), 120 s (**c**) and 200 s (**d**). AFM images of Al_2O_3 morphology for samples treated with varying amounts of fluorine: (**e**) no pretreatment and (**f**) 40 s XeF_2 exposure results in large areas of no oxide deposition; (**g**) 120 s XeF_2 exposure yields a conformal, uniform film and (**h**) 200 s XeF_2 exposure shows pinholes throughout the oxide. Figures adapted with permission from Ref. [39], Copyright Elsevier 2012.

AFM characterization of the Al_2O_3 uniformity was carried out for all the fluorine prefunctionalization conditions. Figure 12e,h shows four representative AFM images of Al_2O_3 deposited on pristine EG (c) and after fluorine prefunctionalization at 40 s (f), 120 s (g) and 200 s (h), respectively. An inhomogeneous Al_2O_3 coverage, especially close to the SiC step edges, was observed on the untreated EG (e) and after 40 s XeF_2 exposure (f). Highly uniform and conformal coverage by the 15 nm Al_2O_3 film was obtained on the fluorine prefunctionalized sample by 120 s XeF_2 exposure, as shown in Figure 12g. This indicated the need of C–F bonding, as observed by XPS (Figure 12c) to promote adhesion of the ALD oxide to the EG film. A slight degradation of Al_2O_3 uniformity with the appearance of small pinholes in the oxide was observed in the sample prefunctionalized by 200 s XeF_2 exposure (Figure 12h). Hence, based on morphological characterization, an optimal window of XeF_2 exposure times from 60 to 180 s was identified to functionalize the EG surface with C–F bonds, resulting in uniform ALD of high-k dielectric films. Noteworthy, an almost unchanged surface roughness was observed comparing the morphology of EG before and after the prefunctionalization under these optimal conditions, indicating that the formation of C-F bonds does not substantially disrupt the EG lattice and planarity. Raman spectroscopy was also performed after the XeF_2 pretreatment and ALD to evaluate the effect of these processes on the underlying EG. In spite of the formation of sp^3 C–F bonds due to the XeF_2 pretreatment, no significant increase in the D/G peaks intensity ratio was observed with increasing the XeF_2 exposure time (i.e., the fluorine percentage on EG surface), suggesting the EG lattice was relatively unperturbed throughout the XeF_2 and ALD process.

6. Open Research Issues and Perspectives

Deposition of ultra-thin and conformal insulators represents a key requirement for the fabrication of devices based on different kinds of graphene. Besides ALD, also physical deposition methods (e.g., evaporation or sputtering) have been explored to this purpose. Although the presence of active sites on graphene surface is not required for these approaches, physical deposition of insulators typically results in graphene damage and/or a reduction of the carrier mobility [56,57]. For this reason, ALD became the preferred approach for insulators deposition on graphene, in spite of the nucleation issues.

Thermal ALD of high-k dielectrics on EG presents some peculiar aspects related to the unique electrical/structural properties of the EG/SiC(0001) system, i.e., the presence of the sp^3 hybridized interfacial BL, responsible for a high n-type doping and compressive strain of the overlying graphene. These properties were shown to be beneficial in enhancing the Al_2O_3 nucleation on 1 L EG, whereas a poorer nucleation was observed on 2 L or few layers EG. Hence, the availability of 100% 1L EG coverage by thermal decomposition of SiC still represents a major challenge for direct ALD on EG.

CVD grown graphene on catalytic metals is another type of graphene widely used for electronics/optoelectronic applications. Typically, this material is transferred on insulating substrate and subsequently processed for devices fabrication. As the thermal ALD process commonly results in an inhomogeneous coverage, seeding layers or prefunctionalization treatments with reactive gas species are commonly employed to promote the ALD growth. However, these processes need to be carefully optimized to minimize structural damage and/or a degradation of the electrical properties (doping and mobility) of underlying graphene.

Recently, Dlubak et al. [58] demonstrated the direct thermal ALD of uniform Al_2O_3 films on monolayer CVD graphene laying on the native metal substrates (Cu and Ni-Au). The enhanced nucleation was ascribed to the wetting transparency of monolayer graphene [59] and to the presence of peculiar polar traps at the graphene/metal interface, which promote the adsorption of water molecules (i.e., the oxygen precursor for the ALD process) on the graphene surface. Clearly, the number of graphene layers is a crucial aspect for this growth mechanism. In fact, the strength of the electrostatic interaction between the water precursor and the polar traps is weakened for multilayer graphene due to a decreased transparency, resulting in an inhomogeneous Al_2O_3 coverage [58]. The direct growth of Al_2O_3 or HfO_2 on CVD graphene residing on the native metal substrate has the advantage of providing a protective layer for graphene, avoiding the direct contact of Gr with the polymeric films typically used for the transfer process [60]. Furthermore, after transfer to an insulating substrate, the oxide layer on graphene can work a gate dielectric for graphene field effect transistors [60]. Although the ALD grown protective layer is beneficial to solve the problems related to polymeric contaminations on graphene surface, the transfer of the oxide/graphene stack from the native metal substrate to the target substrate still remains a critical step.

Most studies reported so far for ALD on graphene concern the deposition of metal-oxides. However, the integration of ultra-thin films of other materials, such as nitrides (including SiN_x, AlN, etc.), hexagonal boron nitride, and transition metal dichalcogenides, TMDs, (such as MoS_2, WS_2, $MoSe_2$ and WSe_2,) with graphene (and specifically with the EG/SiC(0001) system) is attracting an increasing interest for novel electronics/optoelectronics applications [61,62], and ALD can play a key role in this field. To date, MOCVD and MBE are mainly employed for the integration of nitrides with graphene. Although plasma assisted-ALD allows a superior control on the uniformity of SiNx [63] and AlN [64] thin films, the main issue is the plasma-induced damage or doping in graphene. Optimized ALD approaches for the growth of these materials on graphene are envisaged for the next years.

Hexagonal boron nitride (h-BN) is an insulating layered material with lattice structure similar to graphite. Due to the atomically sharp interface formed between h-BN and graphene, it is considered as the ideal insulator to achieve the intrinsic graphene mobility [65]. However, due to its low dielectric constant ($k \approx 4$), the h-BN interfacial layer must be combined with a high-k dielectric overlayer in order to reduce the effective oxide thickness (EOT) for realistic integration in graphene field effect transistors. Recently, the direct ALD growth of a high-k insulator (Y_2O_3) on the sp^2 h-BN surface has been demonstrated, and the deposition mechanism was explained by enhanced adsorption of the Y precursor on h-BN due to the polarization [66]. The challenge in this research field is the large area growth of h-BN on graphene by scalable approaches.

The growth of MoS_2 and other TMDs on EG is currently explored for the realization of 2D materials Van-der Waals heterojunctions on large area [67]. In this context, ALD can represent a valid alternative to the most employed CVD approaches. Dedicated studies on the selection of the precursors and on ALD deposition conditions are expected in the forthcoming years in this emerging research field.

7. Conclusions

In conclusion, the research on ALD of high-k insulators on EG was reviewed, with a focus on the role played by the peculiar electrical/structural properties of the EG/SiC(0001) interface in the nucleation step of the ALD process. Direct thermal ALD of uniform Al_2O_3 thin films on monolayer EG areas was demonstrated to be possible, but achieving monolayer graphene on the entire SiC surface still remains a major challenge. An overview of seeding layers approaches (oxidized metal films, directly deposited metal-oxides and self-assembled organic monolayers) and prefunctionalization treatments (e.g., ozone or fluorine treatments) for the ALD of different high-k materials on EG was provided, considering the impact of these surface preparation processes on the defectivity and electrical properties (doping and carrier mobility) of the underlying EG. Finally, the open scientific issues and the perspectives for ALD growth of alternative insulator/semiconductor films (including AlN, SiN, h-BN and TMDs) on graphene were discussed.

Author Contributions: Writing—original draft preparation, F.G.; writing—review and editing, F.G., E.S., R.L.N., F.R. and R.Y.; supervision, F.G.; funding acquisition, F.G. and F.R. All authors have read and agreed to the published version of the manuscript.

Funding: This work was funded, in part, by MIUR in the framework of the FlagERA-JTC 2015 project GraNitE and the FlagERA-JTC 2019 project ETMOS.

Acknowledgments: The authors want to acknowledge P. Fiorenza, G. Greco, S. Di Franco, I. Deretzis, G. Nicotra and A. La Magna and (CNR-IMM Catania, Italy), A. Armano, M. Cannas, M. Gelardi, and S. Agnello (University of Palermo, Italy), I. Cora and B. Pecz (Hungarian Academy of Sciences, Budapest), I. G. Ivanov and I. Shtepliuk (Linkoping University, Sweden) for useful discussions and the participation in some of the experiments reported in this review paper.

References

1. Novoselov, K.S.; Geim, A.K.; Morozov, S.V.; Jiang, D.; Zhang, Y.; Dubonos, S.V.; Grigorieva, I.V.; Firsov, A.A. Electric Field Effect in Atomically Thin Carbon Films. *Science* **2004**, *306*, 666–669. [CrossRef] [PubMed]
2. Berger, C.; Song, Z.; Li, X.; Wu, X.; Brown, N.; Naud, C.; Mayou, D.; Li, T.; Hass, J.; Marchenkov, A.N.; et al. Electronic confinement and coherence in patterned epitaxial graphene. *Science* **2006**, *312*, 1191–1196. [CrossRef] [PubMed]
3. Emtsev, K.V.; Bostwick, A.; Horn, K.; Jobst, J.; Kellogg, G.L.; Ley, L.; McChesney, J.L.; Ohta, T.; Reshanov, S.A.; Rohrl, J.; et al. Towards wafer-size graphene layers by atmospheric pressure graphitization of SiC(0001). *Nat. Mater.* **2009**, *8*, 203–207. [CrossRef] [PubMed]
4. Virojanadara, C.; Syvajarvi, M.; Yakimova, R.; Johansson, L.I.; Zakharov, A.A.; Balasubramanian, T. Homogeneous large-area graphene layer growth on 6*H*-SiC(0001). *Phys. Rev. B* **2008**, *78*, 245403. [CrossRef]
5. Moon, J.S.; Curtis, D.; Hu, M.; Wong, D.; McGuire, C.; Campbell, P.M.; Jernigan, G.; Tedesco, J.L.; VanMil, B.; Myers-Ward, R.; et al. Epitaxial-Graphene RF Field-Effect Transistors on Si-Face 6H-SiC Substrates. *IEEE Electron Device Lett.* **2009**, *30*, 650–652. [CrossRef]
6. Lin, Y.-M.; Dimitrakopoulos, C.; Jenkins, K.A.; Farmer, D.B.; Chiu, H.-Y.; Grill, A.; Avouris, P. 100-GHz Transistors from Wafer-Scale Epitaxial Graphene. *Science* **2010**, *327*, 662. [CrossRef]
7. Tzalenchuk, A.; Lara-Avila, S.; Kalaboukhov, A.; Paolillo, S.; Syväjärvi, M.; Yakimova, R.; Kazakova, O.; Janssen, T.J.B.M.; Fal'ko, V.; Kubatkin, S. Towards a quantum resistance standard based on epitaxial graphene. *Nat. Nanotechnol.* **2010**, *5*, 186–189. [CrossRef]
8. Pearce, R.; Iakimov, T.; Andersson, M.; Hultman, L.; Lloyd Spetz, A.; Yakimova, R. Epitaxially grown graphene based gas sensors for ultra sensitive NO_2 detection. *Sens. Actuators B* **2011**, *155*, 451–455. [CrossRef]
9. Li, X.; Cai, W.; An, J.; Kim, S.; Nah, J.; Yang, D.; Piner, R.; Velamakanni, A.; Jung, I.; Tutuc, E.; et al. Large-area synthesis of high-quality and uniform graphene films on copper foils. *Science* **2009**, *324*, 1312–1314. [CrossRef]
10. Kang, J.; Shin, D.; Bae, S.; Hong, B.H. Graphene transfer: Key for applications. *Nanoscale* **2012**, *4*, 5527. [CrossRef]
11. Fisichella, G.; Di Franco, S.; Roccaforte, F.; Ravesi, S.; Giannazzo, F. Microscopic mechanisms of graphene electrolytic delamination from metal substrates. *Appl. Phys. Lett.* **2014**, *104*, 233105. [CrossRef]
12. Lin, Y.; Lu, C.; Yeh, C.; Jin, C.; Suenaga, K.; Chiu, P. Graphene Annealing: How Clean Can It Be? *Nano Lett.* **2012**, *12*, 414–419. [CrossRef] [PubMed]
13. Lupina, G.; Kitzmann, J.; Costina, I.; Lukosius, M.; Wenger, C.; Wolff, A.; Vaziri, S.; Ostling, M.; Pasternak, I.; Krajewska, A.; et al. Residual metallic contamination of transferred chemical vapor deposited graphene. *ACS Nano* **2015**, *9*, 4776. [CrossRef] [PubMed]
14. Wu, Y.; Lin, Y.; Bol, A.A.; Jenkins, K.A.; Xia, F.; Farmer, D.B.; Zhu, Y.; Avouris, P. High-Frequency, Scaled Graphene Transistors on Diamond-Like Carbon. *Nature* **2011**, *472*, 74–78. [CrossRef] [PubMed]
15. Mehr, W.; Dabrowski, J.; Scheytt, J.C.; Lippert, G.; Xie, Y.H.; Lemme, M.C.; Ostling, M.; Lupina, G. Vertical Graphene Base Transistor. *IEEE Electron Device Lett.* **2012**, *33*, 691–693. [CrossRef]
16. Giannazzo, F.; Greco, G.; Roccaforte, F.; Sonde, S.S. Vertical Transistors Based on 2D Materials: Status and Prospects. *Crystals* **2018**, *8*, 70. [CrossRef]
17. Giannazzo, F.; Greco, G.; Schilirò, E.; Lo Nigro, R.; Deretzis, I.; La Magna, A.; Roccaforte, F.; Iucolano, F.; Ravesi, S.; Frayssinet, E.; et al. High-Performance Graphene/AlGaN/GaN Schottky Junctions for Hot Electron Transistors. *ACS Appl. Electron. Mater.* **2019**, *1*, 2342–2354. [CrossRef]
18. Pakalla, A.; Putkonen, M. *Handbook of Deposition Technologies for Films and Coatings*, 3rd ed.; Elsevier: Amsterdam, The Netherlands, 2010; pp. 364–391.
19. Wang, X.; Tabakman, S.M.; Dai, H. Atomic Layer Deposition of Metal Oxides on Pristine and Functionalized Graphene. *J. Am. Chem. Soc.* **2008**, *130*, 8152. [CrossRef]
20. Vervuurt, R.H.J.; Karasulu, B.; Verheijen, M.A.; Kessels, W.M.M.; Bol, A.A. Uniform Atomic Layer Deposition of Al_2O_3 on Graphene by Reversible Hydrogen Plasma Functionalization. *Chem. Mater.* **2017**, *29*, 2090. [CrossRef]
21. Schilirò, E.; Lo Nigro, R.; Roccaforte, F.; Deretzis, I.; La Magna, A.; Armano, A.; Agnello, S.; Pecz, B.; Ivanov, I.G.; Giannazzo, F. Seed-Layer-Free Atomic Layer Deposition of Highly Uniform Al_2O_3 Thin Films onto Monolayer Epitaxial Graphene on Silicon Carbide. *Adv. Mater. Interfaces* **2019**, *6*, 1900097. [CrossRef]

22. Lin, Y.-M.; Jenkins, K.A.; Valdes-Garcia, A.; Small, J.P.; Farmer, D.B.; Avouris, P. Operation of Graphene Transistors at Gigahertz Frequencies. *Nano Lett.* **2008**, *9*, 422. [CrossRef]

23. Kim, S.; Nah, J.; Jo, I.; Shahrjerdi, D.; Colombo, L.; Yao, Z.; Tutuc, E.; Banerjee, S.K. Realization of a high mobility dual-gated graphene field-effect transistor with Al_2O_3 dielectric. *Appl. Phys. Lett.* **2009**, *94*, 062107. [CrossRef]

24. Fisichella, G.; Schilirò, E.; Di Franco, S.; Fiorenza, P.; Lo Nigro, R.; Roccaforte, F.; Ravesi, S.; Giannazzo, F. Interface Electrical Properties of Al_2O_3 Thin Films on Graphene Obtained by Atomic Layer Deposition with an in Situ Seedlike Layer. *ACS Appl. Mater. Interfaces* **2017**, *9*, 7761–7771. [CrossRef] [PubMed]

25. Shen, T.; Gu, J.J.; Xu, M.; Wu, Y.Q.; Bolen, M.L.; Capano, M.A.; Engel, L.W.; Ye, P.D. Observation of quantum-Hall effect in gated epitaxial graphene grown on SiC (0001). *Appl. Phys. Lett.* **2009**, *95*, 172105. [CrossRef]

26. Fallahazad, B.; Lee, K.; Lian, G.; Kim, S.; Corbet, C.M.; Ferrer, D.A.; Colombo, L.; Tutuc, E. Scaling of Al_2O_3 dielectric for graphene field-effect transistors. *Appl. Phys. Lett.* **2012**, *100*, 093112. [CrossRef]

27. Meric, I.; Dean, C.R.; Young, A.F.; Baklitskaya, N.; Tremblay, N.J.; Nuckolls, C.; Kim, P.; Shepard, K.L. Channel length scaling in graphene field-effect transistors studied with pulsed current-voltage measurements. *Nano Lett.* **2011**, *11*, 1093. [CrossRef] [PubMed]

28. Farmer, D.B.; Chiu, H.-Y.; Lin, Y.-M.; Jenkins, K.A.; Xia, F.; Avouris, P. Utilization of a Buffered Dielectric to Achieve High Field-Effect Carrier Mobility in Graphene Transistors. *Nano Lett.* **2009**, *9*, 4474. [CrossRef] [PubMed]

29. Lim, T.; Kim, D.; Ju, S. Direct deposition of aluminum oxide gate dielectric on graphene channel using nitrogen plasma treatment. *Appl. Phys. Lett.* **2013**, *113*, 013107. [CrossRef]

30. Schilirò, E.; Lo Nigro, R.; Roccaforte, F.; Giannazzo, F. Recent Advances in Seeded and Seed-Layer-Free Atomic Layer Deposition of High-K Dielectrics on Graphene for Electronics. *C J. Carbon Res.* **2019**, *5*, 53.

31. Lee, B.; Mordi, G.; Kim, M.J.; Chabal, Y.J.; Vogel, E.M.; Wallace, R.M.; Cho, K.J.; Colombo, L.; Kim, J. Characteristics of high-k Al_2O_3 dielectric using ozone-based atomic layer deposition for dual-gated graphene devices. *Appl. Phys. Lett.* **2010**, *97*, 043107. [CrossRef]

32. Lee, B.; Park, S.-Y.; Kim, H.-C.; Cho, K.; Vogel, E.M.; Kim, M.J.; Wallace, R.M.; Kim, J. Conformal Al_2O_3 dielectric layer deposited by atomic layer deposition for graphene-based nanoelectronics. *Appl. Phys. Lett.* **2008**, *92*, 203102. [CrossRef]

33. Vervuurt, R.H.J.; Kessels, W.M.M.; Bol, A.A. Atomic Layer Deposition for Graphene Device Integration. *Adv. Mater. Interfaces* **2017**, *4*, 1700232. [CrossRef]

34. Robinson, J.A.; LaBella, M.; Trumbull, K.A.; Weng, X.; Cavelero, R.; Daniels, T.; Hughes, Z.; Hollander, M.; Fanton, M.; Snyder, D. Epitaxial Graphene Materials Integration: Effects of Dielectric Overlayers on Structural and Electronic Properties. *ACS Nano* **2010**, *4*, 2667–2672. [CrossRef] [PubMed]

35. Hollander, M.J.; LaBella, M.; Hughes, Z.R.; Zhu, M.; Trumbull, K.A.; Cavalero, R.; Snyder, D.W.; Wang, X.; Hwang, E.; Datta, S.; et al. Enhanced Transport and Transistor Performance with Oxide Seeded High-k Gate Dielectrics on Wafer-Scale Epitaxial Graphene. *Nano Lett.* **2011**, *11*, 3601. [CrossRef]

36. Alaboson, J.M.P.; Wang, Q.H.; Emery, J.D.; Lipson, A.L.; Bedzyk, M.J.; Elam, J.W.; Pellin, M.J.; Hersam, M.C. Seeding Atomic Layer Deposition of High-k Dielectrics on Epitaxial Graphene with Organic Self-Assembled Monolayers. *ACS Nano* **2011**, *5*, 5223–5232. [CrossRef]

37. Nath, A.; Kong, B.D.; Koehler, A.D.; Anderson, V.R.; Wheeler, V.D.; Daniels, K.M.; Boyd, A.K.; Cleveland, E.R.; Myers-Ward, R.L.; Gaskill, D.K.; et al. Universal conformal ultrathin dielectrics on epitaxial graphene enabled by a graphene oxide seed layer. *Appl. Phys. Lett.* **2017**, *110*, 013106. [CrossRef]

38. Speck, F.; Ostler, M.; Röhrl, J.; Emtsev, K.V.; Hundhausen, M.; Ley, L.; Seyller, T. Atomic layer deposited aluminum oxide films on graphite and graphene studied by XPS and AFM. *Phys. Status Solidi C* **2010**, *7*, 398–401. [CrossRef]

39. Wheeler, V.; Garces, N.; Nyakiti, L.; Myers-Ward, R.; Jernigan, G.; Culbertson, J.; Eddy, C., Jr.; Gaskill, D.K. Fluorine functionalization of epitaxial graphene for uniform deposition of thin high-k dielectrics. *Carbon* **2012**, *50*, 2307–2314. [CrossRef]

40. Emtsev, K.V.; Speck, F.; Seyller, T.; Ley, L.; Riley, J.D. Interaction, growth, and ordering of epitaxial graphene on SiC{0001} surfaces: A comparative photoelectron spectroscopy study. *Phys. Rev. B* **2008**, *77*, 155303. [CrossRef]

41. Jabakhanji, B.; Camara, N.; Caboni, A.; Consejo, C.; Jouault, B.; Godignon, P.; Camassel, J. Almost Free Standing Graphene on SiC(000-1) and SiC(11-20). *Mater. Sci. Forum* **2012**, *711*, 235–241. [CrossRef]

42. Bouhafs, C.; Zakharov, A.A.; Ivanov, I.G.; Giannazzo, F.; Eriksson, J.; Stanishev, V.; Kühne, P.; Iakimov, T.; Hofmann, T.; Schubert, M.; et al. Multi-scale investigation of interface properties, stacking order and decoupling of few layer graphene on C-face 4H-SiC. *Carbon* **2017**, *116*, 722–732. [CrossRef]

43. Nicotra, G.; Deretzis, I.; Scuderi, M.; Spinella, C.; Longo, P.; Yakimova, R.; Giannazzo, F.; La Magna, A. Interface disorder probed at the atomic scale for graphene grown on the C face of SiC. *Phys. Rev. B* **2015**, *91*, 15541. [CrossRef]

44. Ostler, M.; Deretzis, I.; Mammadov, S.; Giannazzo, F.; Nicotra, G.; Spinella, C.; Seyller, T.; La Magna, A. Direct growth of quasi-free-standing epitaxial graphene on nonpolar SiC surfaces. *Phys. Rev. B* **2013**, *88*, 085408. [CrossRef]

45. Varchon, F.; Feng, R.; Hass, J.; Li, X.; Ngoc Nguyen, B.; Naud, C.; Mallet, P.; Veuillen, J.-Y.; Berger, C.; Conrad, E.H.; et al. Electronic Structure of Epitaxial Graphene Layers on SiC: Effect of the Substrate. *Phys. Rev. Lett.* **2007**, *99*, 126805. [CrossRef]

46. Riedl, C.; Zakharov, A.A.; Starke, U.; Riedl, C. Precise in situ thickness analysis of epitaxial graphene layers on SiC(0001) using low-energy electron diffraction and angle resolved ultraviolet photoelectron spectroscopy. *Appl. Phys. Lett.* **2008**, *93*, 033106. [CrossRef]

47. Sonde, S.; Giannazzo, F.; Raineri, V.; Yakimova, R.; Huntzinger, J.-R.; Tiberj, A.; Camassel, J. Electrical properties of the graphene/4H-SiC (0001) interface probed by scanning current spectroscopy. *J. Phys. Rev. B* **2009**, *80*, 241406. [CrossRef]

48. Vecchio, C.; Sonde, S.; Bongiorno, C.; Rambach, M.; Yakimova, R.; Rimini, E.; Raineri, V.; Giannazzo, F. Nanoscale structural characterization of epitaxial graphene grown on off-axis 4H-SiC (0001). *Nanoscale Res. Lett.* **2011**, *6*, 269. [CrossRef]

49. Giannazzo, F.; Deretzis, I.; La Magna, A.; Roccaforte, F.; Yakimova, R. Electronic transport at monolayer-bilayer junctions in epitaxial graphene on SiC. *Phys. Rev. B* **2012**, *86*, 235422. [CrossRef]

50. Nicotra, G.; Ramasse, Q.M.; Deretzis, I.; La Magna, A.; Spinella, C.; Giannazzo, F. Delaminated Graphene at Silicon Carbide Facets: Atomic Scale Imaging and Spectroscopy. *ACS Nano* **2013**, *7*, 3045–3052. [CrossRef]

51. Ivanov, I.G.; Ul Hassan, J.; Iakimov, T.; Zakharov, A.A.; Yakimova, R.; Janzén, E. Layer number determination in graphene on SiC by reflectance mapping. *Carbon* **2014**, *77*, 492. [CrossRef]

52. Lee, D.S.; Riedl, C.; Krauss, B.; von Klitzing, K.; Starke, U.; Smet, J.H. Raman Spectra of Epitaxial Graphene on SiC and of Epitaxial Graphene Transferred to SiO$_2$. *Nano Lett.* **2008**, *8*, 4320. [CrossRef]

53. Hong, G.; Han, Y.; Schutzius, T.M.; Wang, Y.; Pan, Y.; Hu, M.; Jie, J.; Sharma, C.S.; Muller, U.; Poulikakos, D. On the Mechanism of Hydrophilicity of Graphene. *Nano Lett.* **2016**, *16*, 4447. [CrossRef]

54. Giannazzo, F.; Sonde, S.; Lo Nigro, R.; Rimini, E.; Raineri, V. Mapping the Density of Scattering Centers Limiting the Electron Mean Free Path in Graphene. *Nano Lett.* **2011**, *11*, 4612–4618. [CrossRef]

55. Touhara, H.; Okino, F. Property control of carbon materials by fluorination. *Carbon* **2000**, *38*, 241–267. [CrossRef]

56. Lemme, M.C.; Echtermeyer, T.J.; Baus, M.; Kurz, H. A Graphene Field-Effect Device. *IEEE Electron Device Lett.* **2007**, *28*, 282–284. [CrossRef]

57. Ni, Z.H.; Wang, H.M.; Ma, Y.; Kasim, J.; Wu, Y.H.; Shen, Z.X. Tunable Stress and Controlled Thickness Modification in Graphene by Annealing. *ACS Nano* **2008**, *2*, 1033–1039. [CrossRef]

58. Dlubak, B.; Kidambi, P.R.; Weatherup, R.S.; Hofmann, S.; Robertson, J. Substrate-assisted nucleation of ultra-thin dielectric layers on graphene by atomic layer deposition. *Appl. Phys. Lett.* **2012**, *100*, 173113. [CrossRef]

59. Rafiee, J.; Mi, X.; Gullapalli, H.; Thomas, A.V.; Yavari, F.; Shi, Y.; Ajayan, P.M.; Koratkar, N.A. Wetting transparency of graphene. *Nat. Mater.* **2012**, *11*, 217–222. [CrossRef]

60. Cabrero-Vilatela, A.; Alexander-Webber, J.A.; Sagade, A.A.; Aria, A.I.; Braeuninger-Weimer, P.; Martin, M.-B.; Weatherup, R.S.; Hofmann, S. Atomic layer deposited oxide films as protective interface layers for integrated graphene transfer. *Nanotechnology* **2017**, *28*, 485201. [CrossRef]

61. Giannazzo, F.; Fisichella, G.; Greco, G.; La Magna, A.; Roccaforte, F.; Pecz, B.; Yakimova, R.; Dagher, R.; Michon, A.; Cordier, Y. Graphene integration with nitride semiconductors for high power and high frequency electronics. *Phys. Status Solidi A* **2017**, *214*, 1600460. [CrossRef]

62. Zubair, A.; Nourbakhsh, A.; Hong, J.-Y.; Qi, M.; Song, Y.; Jena, D.; Kong, J.; Dresselhaus, M.; Palacios, T. Hot Electron Transistor with van der Waals Base-Collector Heterojunction and High- Performance GaN Emitter. *Nano Lett.* **2017**, *17*, 3089–3096. [CrossRef] [PubMed]
63. Meng, X.; Byun, Y.-C.; Kim, H.S.; Lee, J.S.; Lucero, A.T.; Cheng, L.; Kim, J. Atomic Layer Deposition of Silicon Nitride Thin Films: A Review of Recent Progress, Challenges, and Outlooks. *Materials* **2016**, *9*, 1007. [CrossRef] [PubMed]
64. Schilirò, E.; Giannazzo, F.; Bongiorno, C.; Di Franco, S.; Greco, G.; Roccaforte, F.; Prystawko, P.; Kruszewski, P.; Leszczyński, M.; Krysko, M.; et al. Structural and electrical properties of AlN thin films on GaN substrates grown by plasma enhanced-Atomic Layer Deposition. *Mater. Sci. Semicond. Process.* **2019**, *97*, 35–39. [CrossRef]
65. Dean, C.R.; Young, A.F.; Meric, I.; Lee, C.; Wang, L.; Sorgenfrei, S.; Watanabe, K.; Taniguchi, T.; Kim, P.; Shepard, K.L.; et al. Boron nitride substrates for high-quality graphene electronics. *Nat. Nanotechnol.* **2010**, *5*, 722. [CrossRef]
66. Takahashi, N.; Watanabe, K.; Taniguchi, T.; Nagashio, K. Atomic layer deposition of Y2O3 on h-BN for a gate stack in graphene FETs. *Nanotechnology* **2015**, *26*, 175708. [CrossRef]
67. Lin, Y.-C.; Ghosh, R.K.; Addou, R.; Lu, N.; Eichfeld, S.M.; Zhu, H.; Li, M.-Y.; Peng, X.; Kim, M.J.; Li, L.-J.; et al. Atomically thin resonant tunnel diodes built from synthetic van der Waals heterostructures. *Nat. Commun.* **2015**, *6*, 7311. [CrossRef]

Review

Electronic and Transport Properties of Epitaxial Graphene on SiC and 3C-SiC/Si: A Review

Aiswarya Pradeepkumar [1], D. Kurt Gaskill [2] and Francesca Iacopi [1,3,*

[1] Faculty of Engineering and IT, University of Technology Sydney, Sydney, New South Wales 2007, Australia; aiswarya.pradeepkumar@uts.edu.au
[2] Institute for Research in Electronics and Applied Physics, University of Maryland, College Park, MD 20742, USA; kurtcapt87@verizon.net
[3] ARC Centre of Excellence in Future Low-Energy Electronics Technologies, Faculty of Engineering and IT, University of Technology Sydney, Sydney, New South Wales 2007, Australia
* Correspondence: francesca.iacopi@uts.edu.au

Received: 29 May 2020; Accepted: 22 June 2020; Published: 24 June 2020

Abstract: The electronic and transport properties of epitaxial graphene are dominated by the interactions the material makes with its surroundings. Based on the transport properties of epitaxial graphene on SiC and 3C-SiC/Si substrates reported in the literature, we emphasize that the graphene interfaces formed between the active material and its environment are of paramount importance, and how interface modifications enable the fine-tuning of the transport properties of graphene. This review provides a renewed attention on the understanding and engineering of epitaxial graphene interfaces for integrated electronics and photonics applications.

Keywords: epitaxial graphene; SiC; 3C-SiC on Si; substrate interaction; carrier concentration; mobility; intercalation; buffer layer; surface functionalization

---

1. Introduction

The properties of graphene as well as other members of the two-dimensional (2D) class of materials differ fundamentally from those of typical electronic materials, which makes this class very attractive for future device applications [1]. A particular appealing property of graphene and other 2D materials is the possibility of dynamically tuning their electrical/electronic properties.

Yet, no matter the application under consideration, which range from light detectors to signal modulators and switches, Hall standards, or chemical sensors, control of transport properties is paramount, and the achievement of the sought-after tunability is not always trivial. The very attractive physical nature of the class (i.e., the large surface to volume ratio [2]) is a double-edged sword as interfaces between the material and its surrounding environment, e.g., substrate, interfaces, adsorbents, and metallization, have a significant effect on conductivity [2–5]. Graphene and 2D materials are made out of surfaces. The accurate control of surfaces and interfaces has historically always been a critical and challenging aspect of semiconductor technology [6,7]. Hence, in order to harness the properties of graphene in electronics, a significant shift in the technological approach needs to be made.

In this review, we explore recent advancements made in the charge transport research on epitaxial graphene (EG) synthesized on SiC and Si substrates as these are necessary ingredients for technological applications. Yet, the properties of EG on these substrates are very different.

For EG on 6H-SiC(0001) (the so-called silicon-terminated surface, Si-face), silicon sublimation results in the formation of EG [8] and the properties are dominated by the $6\sqrt{3} \times 6\sqrt{3}$ rotated 30° reconstruction [9] of the SiC surface. The modification of this reconstruction at the interface after graphene synthesis, often termed the buffer layer, has a profound effect on EG conductivity. In addition, for two or more layers, this reconstruction naturally leads to Bernal stacking [10,11]. For graphene on

SiC(000$\bar{1}$) (aka the carbon-terminated, C-face), the surface reconstructs [9] as (1 × 1), which relaxes the stacking constraint and likely contributes to the observed rotational disorder of the layers [12] and weaker substrate interaction. Moreover, the enhanced Si sublimation rate at screw dislocations likely results in some non-uniformity of thickness [13]. The transport properties of these multilayer graphenes (sometimes termed multilayer epitaxial graphenes, MEG) are complex with at least three conduction channels present [14]. Yet, careful control of nucleation has resulted in one to two layers of EG on the C-face with more easily characterized properties [15].

For graphene on 3C-SiC films on Si surfaces, the synthesis route traditionally employs a modified sublimation approach [16–20] or, more recently, a precipitation method using a metal catalyst alloy with evidence pointing to an epitaxial ordering [21]. For the case of EG on 3C-SiC(100)/Si, no buffer layer is formed, but the current state-of-the-art results in interface oxidation/silicates, which contribute to the conductivity [21]. This is in contrast to EG on 3C-SiC(111)/Si, which does appear to form a buffer layer [17,21] with similarity to that of the buffer layer formed on SiC(0001). However, silicates formation at the EG/3C-SiC interface competes to determine the conductivity.

In the following sections, we begin with a discussion of some basic graphene properties, given in Section 2. The main transport measurements methods are discussed in Section 3. Section 4 provides a focus upon the properties of EG on the SiC and Si substrate components since these dominate the charge transport. Then, we discuss the impact of various situations that modify the conductivity of EG such as ambient adsorbents and metals contacts. Section 5 describes the fine-tuning of the impacts from various surrounding interactions via intercalation, functionalization, and gate control.

2. Electronic Band Structure

2.1. Monolayer Graphene

The electronic properties of graphene were first described by P.R. Wallace in 1947 [22] who used the nearest neighbour tight-binding model (involving only p_z electrons perpendicular to the plane of graphene, i.e., p_z orbitals—hich result in π bands) to approximate the low energy electronic structure of an infinite graphene lattice, which demonstrates the linear dispersion.

$$E(k) = \pm\tau\sqrt{1 + 4cos\left(\frac{\sqrt{3}}{2}ak_x\right)cos\left(\frac{1}{2}ak_y\right) + 4cos^2\left(\frac{1}{2}ak_y\right)} \tag{1}$$

The linear dispersion relation of electrons in single-layer graphene can be written using the strong-coupling approximation by taking into account only the nearest-neighbor interaction (low-energy approximation) as follows.

$$E(\Delta k) = crhv_F\,\Delta k \tag{2}$$

where crh is the reduced Planck's constant, the proportionality or slope is the Fermi velocity, v_F is approximately equal to 10^6 m s^{-1}, and Δk is the momentum relative to the K points of the hexagonal reciprocal unit cell [23]. The band structure of a monolayer graphene (MLG), shown in Figure 1, exhibits the gapless linear dispersion (in k-space known as the Dirac cone) of the π bands at the K-point. This has two important implications. The first is the charge carriers, Dirac fermions, which are massless. The second is semi-metallic behaviour [22,24] because when the Fermi energy, E_F, is located where the dispersion converges to a point (the Dirac point, E_D), the density of states is zero, and electronic conduction is possible only through thermally excited electrons [25]. The linear dispersion of monolayer graphene results in several extraordinary electronic properties such as an anomalous quantum Hall effect with Berry's phase of π [8,26–29].

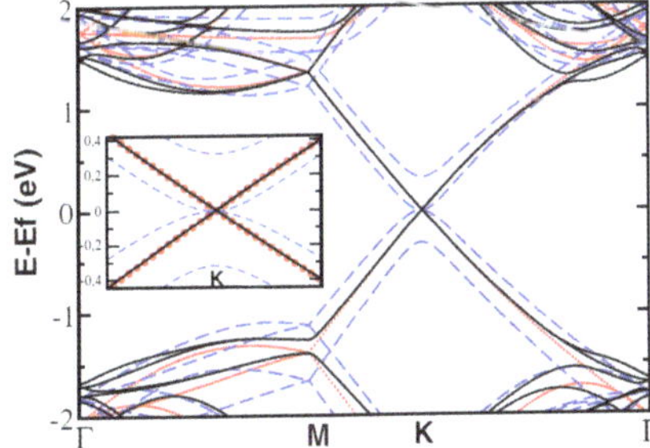

Figure 1. Band structure for a monolayer graphene (dotted line), a bilayer graphene with AB stacking (dashed line), and R30/R2$^+$ rotational fault pair (single line). Band structure at the *K*-point is given in the inset. Reprinted with permission from Reference [23]. Copyright (2008) by American Physical Society.

The tight-binding model oversimplifies the situation of EG on SiC(0001) as a buffer layer exists between SiC and EG. Mattausch and Pankratov [30] and Varchon et al. [31] performed a first-principles calculation, which included the buffer layer and determined the E_F was about 450 meV above the E_D due to doping of the graphene layer by charge transfer from the buffer layer. This is in agreement with angle-resolved photoelectron spectroscopy (ARPES) measurements [26,32–34]. Sprinkle et al. [35] found the average Fermi velocity, obtained from the slope of $E(\Delta k)$ as $v_F = 1.0\ (\pm 0.05) \times 10^6$ ms^{-1} for energies down to ~500 meV above E_D [19,36,37]. This value of v_F is larger than the v_F for bulk graphite ($v_F = 0.86 \times 10^6$ ms^{-1}) [38].

Ouerghi et al. measured the band structure of the monolayer EG/3C-SiC(111) using ARPES at room temperature [19] (see Figure 2) and found the E_F. Using the linear dispersion, the charge carrier concentration, *n*, of monolayer graphene can be obtained as:

$$n = (E_F - E_D)^2/(\pi)(hv_F)^2 \tag{3}$$

where *h* is the Planck's constant, $v_F = 1.1 \times 10^6$ m s^{-1}, and E_F is 500 meV. This resulted in $n \sim 2 \times 10^{13}$ cm^{-2}.

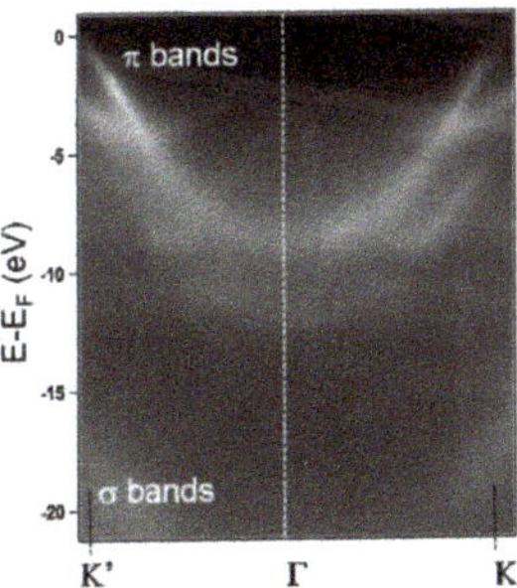

Figure 2. Electronic structure of monolayer EG on 3C-SiC(111) by ARPES. Reprinted with permission from Reference [19]. Copyright (2010) by American Physical Society.

The influence of graphene-substrate interaction on v_F of monolayer epitaxial graphene was estimated by Davydov [39] in which the substrate interaction reduces the v_F of electrons.

2.2. Bilayer Graphene

Bilayer graphene (BLG) synthesized on SiC(0001) are generally reported to be AB (Bernal) stacked with a 30° rotation with respect to the substrate [11]. In BLG, the charge carriers tunnel quantum mechanically between the two layers, which causes a band dispersion that is nearly parabolic [40] (Figure 1) with an effective mass (m^*) of about 0.033 m_e (electron mass) [41].

$$E_{k\pm} = \pm v_F^2 k^2 t^{-1}$$

(4)

$E_{k\pm}$ describes the two bands with energies E_+ and E_- that lies at the K point, $t \sim 3$ eV is the energy of electron transition between the nearest neighbors spaced by a = 0.142 nm (lattice constant).

In this case, the v_F depends on the Fermi momentum (p_F) divided by m^* as p_F/m^* [42]. For higher energies corresponding to doping above 5×10^{12} cm^{-2}, the linear dependence is a good approximation [41]. Ohta et al. performed an early investigation of the band structure properties of BLG on SiC(0001) using ARPES, and found E_F to be 400 meV [34]. De Heer et al. reports v_F to be 0.7–0.8×10^6 ms^{-1} for BLG, which is 20% to 30% smaller than that of MLG with an E_F at ~300 meV above the Dirac point [31]. However, recent measurements put v_F of MLG closer to 1.2×10^6 ms^{-1} [43] with an E_F of ~350 meV. Bernal stacked layers exhibit Berry's phase of 2π [40] as confirmed by the quantum Hall effect [44].

Charge transport in BLG, thus, involves massive chiral and parabolic dispersion carriers [42] (see Figure 1 with a zero band gap). The effect of the buffer layer is not included in contrast to the massless chiral, linear-dispersion carrier system for MLG.

The band structure of BLG is sensitive to the lattice symmetry [34]. When the graphene layers of a BLG are made asymmetric, a band gap forms between the low-energy bands at the former Dirac point [34,40] (see Figure 3). This means the band gap can be varied by means of an external transverse electrical field to control the density of electrons. The gap can be varied from 0 to 0.3 eV by changing the doping level between the two layers or by an external gate control [34,45]. Thus, the BLG transitions from a semi-metal to an insulator [34]. An externally controlled symmetry breaking also changes the electronic conductivity and forms a BLG switch [34]. For this reason, digital electronic applications have been envisioned [25].

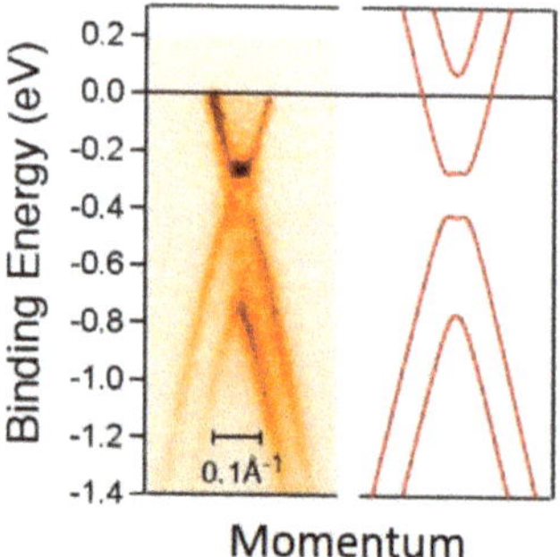

Figure 3. Band structure of BLG on SiC(0001) from ARPES and theoretical (solid lines). Reprinted with permission from Reference [34]. Copyright (2006) American Association for the Advancement of Science.

2.3. Turbostratic Multilayer Graphene

Turbostratic multilayer graphene layers are rotationally disordered and, thus, electrically decoupled. Epitaxial graphene formed on SiC(000$\bar{1}$) by thermal decomposition on both (0001) and SiC(000$\bar{1}$) by solid-phase epitaxy and on 3C-SiC/Si via the alloy mediated graphitization is reported to be turbostratic [21,46,47]. The rotational disorder in EG/SiC(000$\bar{1}$) was shown by low-energy electron diffraction (LEED) investigations in which an azimuthal diffraction pattern was found, and the strong intensity modulation in the bands denoted the rotational orientation [10,12,48]. In the case of EG/3C-SiC, the presence of turbostratic in-plane modes in the Raman spectroscopy indicated the rotational disorder [21]. With rotational disorder, the linear dispersion is recovered in the vicinity of the K-points. The ab initio calculations show that the dispersion is nearly identical to that of an isolated MLG linear dispersion (see Figure 1), and predicts v_F to be the same as that of MLG [12,22,31].

Sprinkle et al. measured the linear dispersion using ARPES, as shown in Figure 4 [35]. In the ARPES data, the linear dispersions of adjacent decoupled layers can be readily seen as well as the n-type conductivity, where E_F is as low as ~14 meV, and the carrier density is ~10^{11} cm^{-2}. Furthermore, the v_F of MLG is close to the v_F of MEG formed by thermal decomposition of the SiC(000$\overline{1}$) obtained from infra-red measurements (1.02 ($\pm$0.01) $\times$ 10^6 ms^{-1}) [49] and scanning tunneling spectroscopy (1.07 ($\pm$0.01) $\times$ 10^6 ms^{-1}) [50].

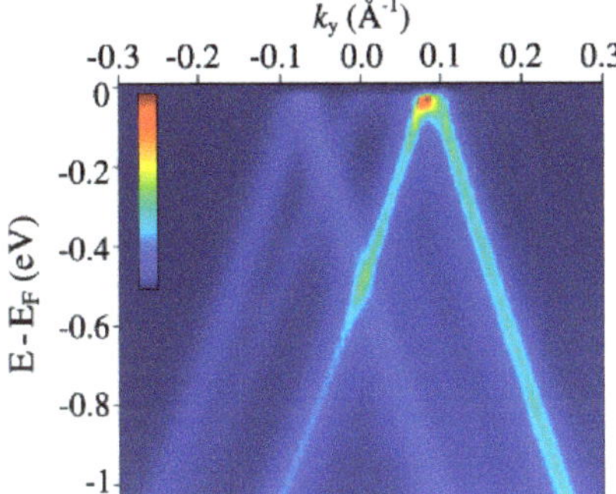

Figure 4. Band structure of EG on (000$\overline{1}$) 6H-SiC with 11 epi-layers obtained from ARPES. Two linear Dirac cones are observed. Reprinted with permission from Reference [35]. Copyright (2009) by American Physical Society.

In the next sections, we will discuss the electronic and transport properties of EG on SiC and 3C-SiC substrates with respect to the substrate, ambient, and metal contact interactions. Section 3 will discuss the different measurement methods used for the charge carrier transport study.

3. Transport Measurement Methods

Knowledge of the transport properties of graphene such as sheet carrier concentration, mobility, and sheet resistance and the factors affecting them are crucial for any applications of the graphene into practical devices. These include Hall effect measurements using Hall bar devices and van der Pauw (vdP) structures, Field-effect measurements using gated devices, Raman spectroscopy (carrier concentration from G peak shift), and ARPES (discussed in Section 2.1).

3.1. Hall Bar Devices

The first transport measurements were performed using Hall bar structures patterned on 6H-SiC(0001) by Berger et al. [4]. In this case, low mobility values were obtained in most of the samples, which may have been limited by the substrate step edges. Yet, the samples clearly demonstrated 2D electron gas properties and most of the important transport features of epitaxial graphene [4].

At the step edges, a second graphene layer is often formed, which is associated with step bunching and results in non-uniform graphene thickness [17] and unreliable transport measurement results [51]. S-H Ji et al. demonstrated that a strong impact of substrate steps degrades the transport properties of epitaxial graphene on SiC using scanning tunneling microscopy-based localized transport measurements [51]. Yakes et al. reported somewhat similar results noting that conductivity anisotropy in graphene was associated with substrate steps. The observed results were explained using a model where a charge build-up at the substrate steps leads to scattering [52]. For these reasons, Hall bar structures are often patterned to be parallel or perpendicular to steps (see Figure 5) to mitigate the effects of steps, and this approach was used for quantum Hall effect [53] devices. We note that Kruskopf and co-workers developed an approach to form EG on SiC(0001) that is essentially free of steps [54] and demonstrated excellent quantum Hall effect results.

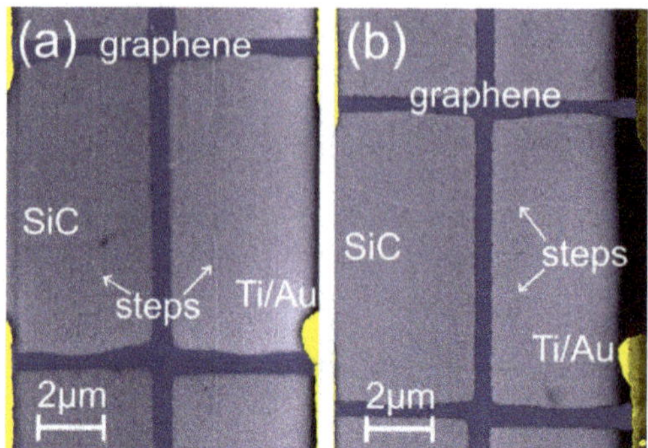

Figure 5. Scanning electron microscopy images of Hall bars fabricated on MLG (**a**) on a flat terrace and (**b**) perpendicular to the step edges on 6H-SiC. Reprinted with permission from Reference [55]. Copyright (2011) Elsevier Ltd.

3.2. Van der Pauw and Hall Bar Structures

Transport measurements from the van der Pauw method deliver large scale measurements [21,55,56] and provide complete information on carrier transport in EG. vdP and Hall bar-based measurements are limited by thickness inhomogeneities and discontinuities in the epitaxial films [21,57]. Jobst et al. demonstrated that the transport properties obtained from Hall bar measurements are almost the same as those from van der Pauw measurements, as shown in Figure 6 [55].

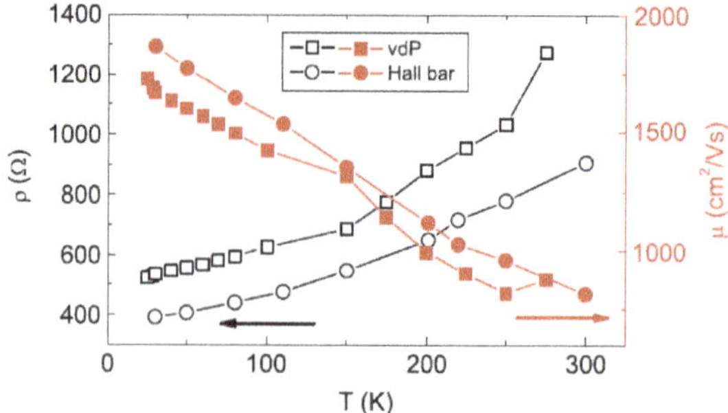

Figure 6. Resistivity, ρ, and mobility, μ, as a function of temperature, T for MLG on SiC, indicating that the measured properties are almost similar for vdP (squares) and Hall bar (circles) structures. Reprinted with permission from Reference [55]. Copyright (2011) Elsevier Ltd.

3.3. Field-Effect Measurements

The electric field-effect from an externally applied voltage to a gate enables the study of the extraordinary electrical properties of graphene [29,58]. The first observation of the electric field-effect in graphene (using mechanically exfoliated graphene onto SiO_2/Si) was reported by K.S. Novoselov and A.K. Geim [29], where the charge transport was switched between electron and hole gases via the gate voltage. Berger et al. performed the first transport studies of EG on SiC(0001) and modulated the electron density via electrostatic gating, which demonstrated the 2D electron gas properties [4]. The deleterious anisotropy effects of step bunching on SiC(0001) on gated devices has been discussed by Lin et al. [59]. In the case of graphene on SiC(000$\bar{1}$), the non-uniform thickness limits gate device applications [60].

In addition, the transport properties from field-effect measurement are dependent on geometry and electrostatics and are affected by the substrate [61]. This transport measurement method does not permit a systematic analysis of charge scattering within the system [21].

3.4. Raman Spectroscopy

The values of phonon frequencies of EG on SiC depend on the mechanical strain and charge transfer between SiC and graphene [62]. The Raman G-band and the 2D band involve phonons at the $K + \Delta K$ points in the Brillouin zone [62–64]. The influence of the doping on the Raman bands has been studied by Das et al. [65] and Rohrl et al. [62] using a gate voltage controlled graphene transferred onto SiO_2/Si. G-peak frequency showed an upshift up to 20 cm^{-1} when the graphene is n-type doped at 4×10^{13} cm^{-2}. The influence of doping on the 2D-peak shift was shown to be weak and is ~10–30% compared to the G-peak shift (3–5 cm^{-1}) [66] (see Figure 7). The value of 2D to G peak intensity was determined by the carrier concentration [65] and indicates the graphene doping level. Mueller et al. [67] reported a method to extract the value of doping from the Raman G and 2D modes of graphene formed on any arbitrary substrates. Furthermore, Verhagen et al. [68] estimated the temperature dependent doping of monolayer and bilayer graphene on SiO_2/Si using Raman spectral mapping between 10 K and 300 K. More recently, in Reference [69], Shtepliuk et al. estimated the thickness dependent electron doping of EG/SiC by silver films from the red-shift and broadening of the 2D band. The same research group studied the temperature and time dependency of the H-intercalation on the p-type doping of quasi-free standing monolayer graphene on EG/SiC via micro-Raman spectroscopy in Reference [70].

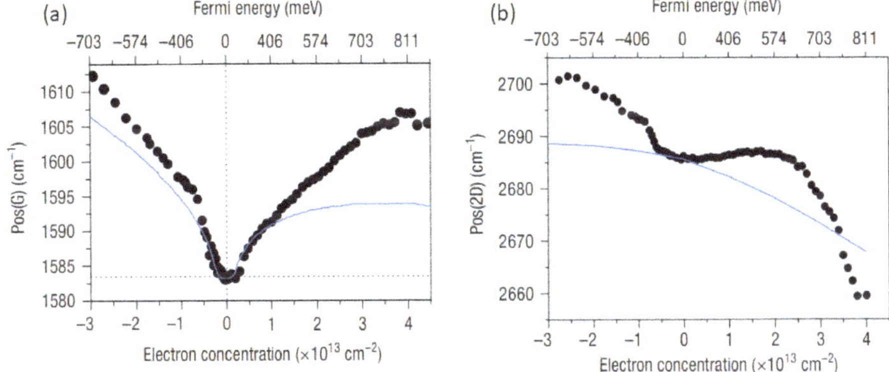

Figure 7. Position of the (**a**) G peak and (**b**) 2D peak as a function of electron and hole doping in graphene on SiO_2. The solid blue line in (**a**) is the predicted non-adiabatic trend from Reference [71]. The solid line in (**b**) is from density functional theory calculation. Reprinted with permission from Reference [65]. Copyright (2008) Springer Nature.

4. Electronic and Transport Properties: Effect of Interactions with the Surroundings

Table 1 shows a summary of electronic and transport properties of as-grown EG on SiC(0001), SiC(000$\bar{1}$), and 3C-SiC/Si substrates showing the effect of EG-substrate and EG-ambient interactions as well as the properties fine-tuned via intercalation, functionalization, and gate control. These properties are discussed throughout the following sections.

Table 1. Summary of electronic and transport properties of as-grown epitaxial graphene on SiC(0001), SiC(000$\bar{1}$), and 3C-SiC/Si substrates indicating the effect of substrate and ambient and fine-tuning of properties using intercalation, functionalization, and gate control.

Substrate	Growth Process	No. of EG Layers	Transport Properties at 300 K						Reference
			Fermi Level, E_F (meV)	n/p	n (cm^{-2})	μ (cm$^2 \cdot$V$^{-1} \cdot$s^{-1})	R_{sh} ($\Omega/\square$)	Measurement Technique	
			EG-Substrate Interaction						
6H-SiC(0001)	Thermal decomposition	1	350–500	n	10^{13}	1000	-	Hall bar	[26,43]
6H-SiC(0001)	Thermal decomposition	2	300–400	n	5×10^{12}	1000	-	Hall bar	[31,34]
6H-/4H-SiC(0001)	Thermal decomposition	Multi	-	n/p	10^{13} –10^{14}	1000–3000		Hall bar	[56]
6H-/4H-SiC(000$\bar{1}$)	Thermal decomposition	Multi (top) (mid) (first)	-	p p n	2×10^{13} 5×10^{11} 2×10^{13}	1500 19,300 12,400	-	Hall bar	[14]
4H-SiC(000$\bar{1}$)	Thermal decomposition	First layer	350	n	9×10^{12}	~1500	-	Optical Hall effect	[72,73]
4H-SiC(000$\bar{1}$)	Thermal decomposition	Multi	-	p	10^{12}	15,000	-	Hall bar	[15]
3C-SiC(100) 3C-SiC(111)	Alloy mediated	3–7	450	p	3×10^{13} 10^{12}	<80 330	~10k	vdP	[21]
			EG-Ambient Interaction						
4H-SiC(0001)	Thermal decomposition	1		p	10^{12}	~1550	-	Optical Hall effect	[74]
			Fine-Tuning of Transport Properties						
			H-Intercalation						
4H-SiC(0001) 6H-SiC(0001)	Thermal decomposition	Quasi-free-standing monolayer	340 300	p	9×10^{12} 6×10^{12}	-	-	ARPES	[75]

Table 1. *Cont.*

Substrate	Growth Process	No. of EG Layers	Fermi Level, E_F (meV)	n/p	n (cm^{-2})	μ (cm^2·V^{-1}·s^{-1})	R_{sh} ($\Omega/\square$)	Measurement Technique	Reference
								Transport Properties at 300 K	
3C-SiC(111)	Thermal decomposition	Quasi-free-standing monolayer	-	n	10^{12}	-	-	APRES	[76]
6H-SiC(0001)	Thermal decomposition	Quasi-free-standing bilayer	230	p	10^{12}	-	-	APRES	[77]
3C-SiC(111)	Alloy mediated	3–7	320	p	3×10^{12}	350	6k	vdP, density functional theory	[21]
Oxygen Intercalation									
6H-SiC(0001)	Thermal decomposition	1	300 200	n	5×10^{12} 2×10^{12}	-	-	ARPES (partial intercalation)	[78]
Magnesium Intercalation									
6H-SiC(0001)	Thermal decomposition	Quasi-free-standing bilayer	720	n	2×10^{14}	-	-	APRES, density functional theory	[43]
Functionalization									
F4-TCNQ/Si-face SiC	Thermal decomposition	1	10	n	5×10^{10}	29000	-	Hall bar (25 K)	[55,79]
Top-gate Graphene Field-Effect Transistor									
SiC(0001) SiC(000$\bar{1}$)	Thermal decomposition	Multilayer	-	n/p	-	600–1200 (0001) 500–3000 (0001)	-	Field-effect transistor	[60]
3C-SiC(111)/Si(111)	Thermal decomposition	-	-	n/p	-	175 (n) 285 (p)	-	Field-effect transistor	[80]

4.1. Epitaxial Graphene-Substrate Interaction

4.1.1. Induced Pseudo-Charge Due to Substrate Polarization Effect

As described by the ARPES results summarized in Section 2, EG on SiC(0001) exhibits an E_F is quite large, which indicates considerable n-type doping. Although some density functional theory (DFT) calculations imply this result, the physical reasons have been clarified by the experimental work of Ristein et al. and Mammadov et al. [75,81]. The doping originates from three effects and, in some cases, existing simultaneously: (i) polarization induced pseudo-charge due to the hexagonal nature of 6H-SiC and 4H-SiC substrates, (ii) surface states associated with C and Si dangling bonds overlaid on the broad density of states (DOS) of the buffer layer, which act as donor states (part of the Kopylov model [82]), and (iii) the effect of space-charge layer in (doped) SiC or Si substrates due to band bending at the interface. Figure 8 illustrates these effects. The referenced works [75,81] demonstrate these reasons are in reasonable agreement with the experiment, especially for hydrogen intercalated graphene (see Section 5.1).

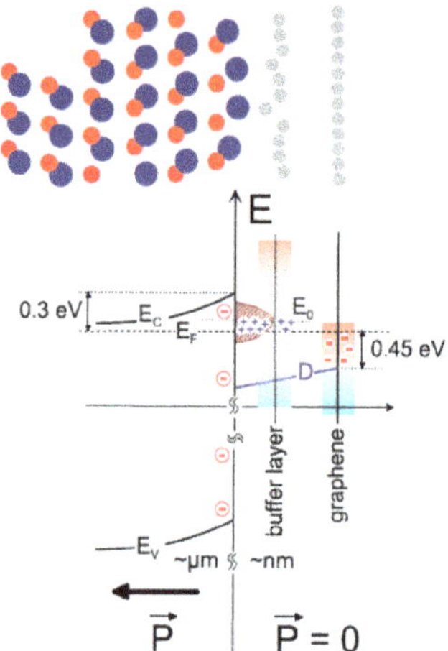

Figure 8. Sketch and band diagram for monolayer EG on SiC(0001). Blue and red circles indicate Si and C atoms, respectively. The polarization is shown at the bottom. The negative polarization related to the discontinuity of the interface polarization is a pseudo charge (marked in circles). D indicates the electrostatic potential between the EG and SiC. The band bending details have been taken from Reference [83]. Adapted with permission from Reference [81]. Copyright (2012) American Physical Society.

The implications of this understanding are significant. (1) Preparing EG on a defect-free interface of SiC (000$\bar{1}$) should result in n-type doping. (2) EG on 3C-SiC should lead to n-doped graphene layers—modified only by the induced effects of substrate donor states. (3) Varying the hexagonality of the substrate will change the doping. This has been successfully tested using 4H-SiC and 6H-SiC [75].

The spontaneous polarization has been reported as a bulk property of hexagonal semiconductor compounds, which leads to a polarization charge on the polar surface of SiC, independent to any interface formation [75,81]. This is induced at the inversion of the stacking sequence of the hexagonal double layers while it is absent in the 3C-SiC polytype due to the symmetry [75]. The spontaneous polarization generates a pseudo acceptor layer with a pseudo charge density (Figure 8) depending on the hexagonal polytype [75]. Ristein et al. demonstrated that the sign and magnitude of SiC polarization agree with the charge concentration of H-intercalated graphene on SiC(0001) [81]. The hexagonality of the SiC affects the polarization proportionally, i.e., for the 4H-SiC(0001) the polarization is 6/4 times larger than that for the 6H-SiC(0001) [75,81] and, as mentioned above, has been shown experimentally via the H-intercalation (Section 5.1).

4.1.2. Impact of Growth on SiC (0001)

Thermal decomposition of SiC in vacuum or argon atmosphere is the well-established process for producing graphene on SiC [5,56,75,84]. Electronic and transport properties of EG on SiC depends upon the type of surface termination of SiC (Si-face or C-face). Although both Si and C sublimate, the Si flux is dominant. The differences between the graphitic layer grown on the Si-face (0001) and the C-face (000$\bar{1}$) were first reported by Bommel et al. [85]

Graphene grown on SiC(0001) possesses a carbon-rich amorphous interfacial layer ($6\sqrt{3} \times 6\sqrt{3}$) rotated 30° [86] known as the buffer layer in between the SiC and graphene [31]. The buffer layer forms strong covalent bonds with the SiC. The electronic structure of the buffer layer shows a large gap and an E_F pinned by a state with a small dispersion close to the conduction band (see Figure 9) [31]. These states are related to the dangling bonds in the buffer layer. When more than one carbon layer is present, the graphene-related dispersions are recovered.

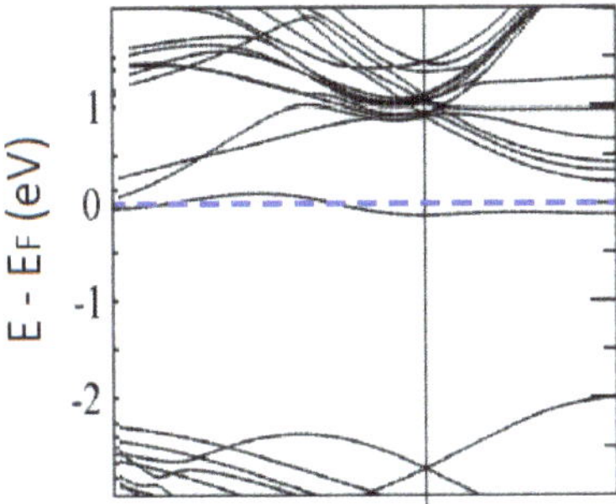

Figure 9. Dispersion curves for a buffer layer. Adapted with permission from Reference [31]. Copyright (2007) American Physical Society.

Emtsev et al. used ARPES measurements to study the band structure changes starting from the $6\sqrt{3}$ surface of SiC. The interaction of the buffer layer with the $6\sqrt{3}$ surface results in a σ-band instead of a π-band. Continued sublimation converts the buffer layer into graphene with π-band [86] and simultaneously creates a new buffer layer underneath. As-grown MLG on Si-face SiC is generally reported to be electron-doped with a carrier concentration of ~1 × 10^{13} cm^{-2} [26,37,82], 300 K mobility of ~1000 cm$^2 \cdot$V$^{-1} \cdot$s^{-1}, and a 25 K mobility of ~2000 cm$^2 \cdot$V$^{-1} \cdot$s^{-1} [26] [55]. The n-type conductivity is due to the charge transfer from donor-like states at the EG/SiC interface, i.e., the buffer layer, that overcompensates the spontaneous polarization of SiC [81], as discussed in Section 4.1.1.

The band structure for BLG on SiC(0001) was calculated using DFT by Varchon et al. [31] and is in agreement with the ARPES results by Ohta et al. [33,34]. Additional details are in Section 2.2. The E_F is ~250–400 meV above the Dirac point (n-type conductivity), which is slightly lower than that for epitaxial monolayers. This is likely due to charge transfer from the buffer layer [31], whose cause likely has some similarity with the effects discussed in Section 4.1.1. The carrier concentration ≈ 5 × 10^{12} cm^{-2} [65] and the mobility is ~1000 cm$^2 \cdot$V$^{-1} \cdot$s^{-1} [45]. The lower E_F and doping level of BLG may indicate a weaker substrate interaction as reported by Ohta et al. [34]

4.1.3. Impact of Growth on SiC (000$\bar{1}$)

Unlike the graphene on SiC (0001), graphene formation on SiC (000$\bar{1}$) does not involve a buffer layer at the EG-SiC interface. This is because a different surface reconstruction occurs. Seurbet et al. [87] analyzed the atomic structure of the (2 × 2) reconstruction on (000$\bar{1}$) SiC using an in situ prepared samples and quantitative LEED intensity analysis. In this case, $\frac{3}{4}$ of the carbon atoms on the $(000\bar{1})$ surface plane are bonded to a silicon adatom in a hollow site with two dangling bonds (Si adatom and C rest atom) [5].

Using first-principles, Varchon et al. calculated the band structure incorporating the (2×2) surface reconstruction (see Figure 10a) [5] and claimed that the electronegativity difference between silicon and carbon induces a charge transfer from the Si adatom to the C rest atom that forms surface states in the electronic band structure EG on $SiC(000\bar{1})$ (see Figure 10b) [31]. This results in pinning of the Fermi energy, which makes the samples n-type. Yet, Ristein et al. and Mammadov et al. argue the situation is more complex due to the hexagonal nature of the substrate (see Section 4.1.1) even though n-type behaviour is still predicted. This situation is clarified by detailed experimental work, which will be described below.

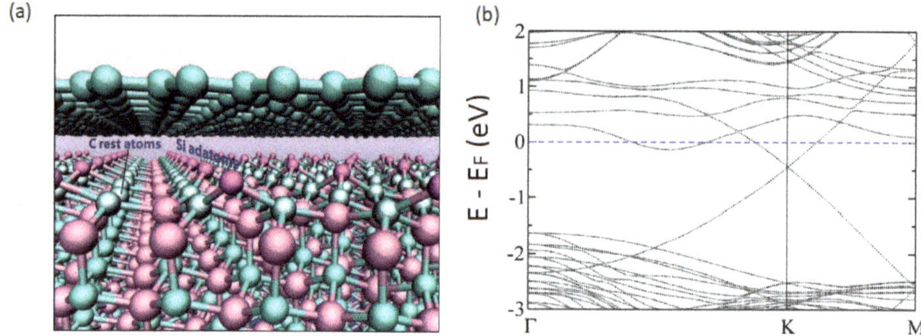

Figure 10. (a) $EG/SiC(000\bar{1})$ interface with the (2×2) C surface reconstruction. Si adatoms and C rest atoms are marked. Adapted with permission from Reference [5]. Copyright (2012) AIP Publishing. (b) Electronic band structure of EG on $SiC(000\bar{1}\ )$. Adapted with permission from Reference [31]. Copyright (2007) American Physical Society.

The graphene layers on the C-face are thicker and randomly rotated against each other as well as with respect to the substrate [86,88] and, thus, exhibit weaker interactions with the substrate [48,86]. This unusual rotational stacking causes multilayers of graphene on the C-face to have an electronic structure similar to that of the monolayer graphene [8,35,88] with a well-defined Dirac cone (linear dispersion) near the charge neutrality point [22]. Shubnikov-de Haas oscillations were observed using Hall bars, which indicated that the transport layer has a Berry phase of π, similar to that of a single layer graphene. This signified that it is electronically decoupled from the layer above it [89]. Additional evidence for the decoupling is found in ARPES data of the multilayer EG (MEG) stack, which reveals the decoupled nature of the layers [35,48] in contrast to ARPES results on Bernal BLG on Si-face, where Dirac cones remain unperturbed and distinct from one another [90]. Infrared spectroscopy measurements shows E_F ~8 meV, doping ~10^{10} cm^{-2}, and low magnetic field mobility ~10^6 cm$^2 \cdot$V$^{-1} \cdot$s^{-1} [91].

Room temperature transport measurements reported on these MEG on C-face SiC samples are typically p-type [5] with sheet density of ~10^{13}–10^{14} cm^{-2} and mobility of ~1000–3000 cm$^2 \cdot$V$^{-1} \cdot$s^{-1} even though a wide range of values and even n-type measurements have been reported [14,56,72–74,92]. This seems to run counter to the n-type predictions noted in Section 4.1.1, which we will address next.

This difficulty of measured p-type properties vs. the n-type prediction situation is explained by considering the work of Lin et al. [14] who performed variable magnetic field Hall measurements on a series of samples. The data was best fit to a model where three types of transport regions exist in the layers of the sample: the layers closest to the substrate (substrate interaction), the interior layers (nearly neutral), and then the outer layers (ambient interaction). Each region has distinct sheet density and mobility. The epitaxial layer that lies closest to the SiC is highly electron-doped with high conductivity

(10^{12} cm^{-2}, 12,400 cm^2·V^{-1}·s^{-1}). This is consistent with the models of Ristein and Mammadov [75,81]. The model is also supported by an ultrafast optical spectroscopy measurement that resulted in a doping density of 9×10^{12} cm^{-2} corresponding to a Fermi level of ~350 meV above the Dirac point for the layer close to the substrate [72].

The second transport region is nearly intrinsic (p-type, 5×10^{11} cm^{-2}, 20,000 cm^2·V^{-1}·s^{-1}) due to charge screening. Additional support for this can be found, for example, from Landau level spectroscopy where the results are expected to be dominated by sample regions having high mobility [90]. Measurements of MEG have resulted in exceptionally high room temperature mobilities (>200,000 cm^2·V^{-1}·s^{-1}) consistent with nearly intrinsic doping [48,50,93,94].

The third transport region is the heavily p-doped outermost layers (10^{13} cm^{-2}, 1500 cm^2·V^{-1}·s^{-1}) likely due to environmental dopants [14] related to the ambient temperature [90]. Sidorov et al. determined that the environmental factors cause a p-type behaviour in ambient exposed monolayer and multilayer EG on SiC(000$\bar{1}$) [5] (a detailed discussion on the influence of ambient temperature on the transport of EG is given in Section 4.2). Lebedev et al. remarked that the outer layers of graphene, which are p-type, play a protecting role for the interior graphene layers, preventing influence from the atmosphere [92]. Sidorov et al. also reported that the natural conductivity state of MEG without the influence of ambient is n-type [5,84], which is also consistent with the substrate hexagonality argument. The difficulty in understanding the range of doping types reported by others is clarified by Lin since many samples analyzed using a simple one-layer model will tend to be p-type even though n-type results can occur depending upon overall thickness and uniformity. Hence, the variations in reported results are likely due to researchers assuming a simple model of uniform charge density and mobility when analyzing Hall data coupled with sample-to-sample variability.

A significant advancement in the transport of EG on 4H-SiC(000$\bar{1}$) was achieved when Wu et al. [15] synthesized a monolayer EG on the C-face, which was p-doped ~1.27×10^{12} cm^{-2} (likely due to environmental doping) with mobility of 20,000 cm^2·V^{-1}·s^{-1} at 4 K and ~15,000 cm^2·V^{-1}·s^{-1} at 300 K [95]. The same group has also demonstrated that the scattering effects due to the substrate (i.e., charge impurity, electron-phonon) are weak. This is related to the absence of a buffer layer, a weak EG-substrate interaction, and no induced charges into the graphene or scattering centres [56,96]. The values of MLG on 4H-SiC(000$\bar{1}$) are an order of magnitude larger than MLG on SiC(0001) and are in line with the values from exfoliated graphene on SiO$_2$ [97,98], even though later studies reported two-fold larger mobility of ~10^6 cm^2·V^{-1}·s^{-1} at a carrier concentration of ~10^8 cm^{-2} and E_F within 1 meV for suspended graphene devices [99–101].

4.1.4. Impact of Growth on (100) and (111) 3C-SiC/Si

Synthesis of EG on Si substrates received attention primarily due to its compatibility with current micromachining technology and processes, low production cost, and ability to synthesize on large areas of substrates [21,66]. Thermal decomposition of 3C-SiC was widely adopted to form graphene on silicon substrates using 3C-SiC(111) and 3C-SiC(100) pseudo-substrates [16,18,20,66,80]. Yet, the formation EG on 3C-SiC via thermal decomposition has a limitation of inconsistent graphene coverage [102]. Pradeepkumar et al. [21] have overcome the coverage issue of graphene on 3C-SiC by using a liquid-phase Ni/Cu alloy-mediated graphene synthesis approach.

Similar to EG formed on SiC(0001), graphitization of (111) type 3C-SiC involves a buffer layer at the EG/3C-SiC interface [18]. Figure 11 shows the presence of the buffer layer in EG/3C-SiC(111) using X-ray photoelectron spectroscopy (XPS) measurements [21], which is also demonstrated in References [103] and [17]. The existence of the buffer layer in the EG/3C-SiC(111) substrate has also been proven by the transmission electron microscopy (TEM) measurements by Fukidome et al. [104]. Figure 11 also indicates that the EG formed on 3C-SiC(100) does not possess a buffer layer [21].

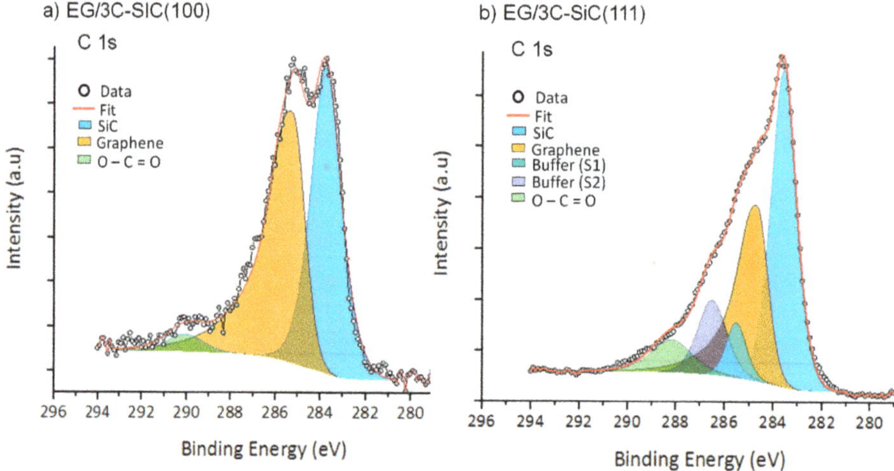

Figure 11. XPS C 1s core-level spectrum for (**a**) EG/3C-SiC (100) and (**b**) EG/3C-SiC (111). Buffer layer components are visible for EG/3C-SiC (111). Reprinted with permission from Reference [21]. Copyright (2020) American Chemical Society.

Graphene layers formed by thermal decomposition on 3C-SiC(111) are Bernal stacked as observed from scanning tunneling microscopy (STM) with a $(6\sqrt{3} \times 6\sqrt{3})$ rotated 30° construction at the interface similar to the EG on SiC(0001) [86]. On the other hand, the epi-layers on 3C-SiC(100) are rotated ±15° with respect to the substrate from LEED measurements by Ouerghi et al. [105]. Graphene formed by alloy-mediated graphitization on 3C-SiC(100) and 3C-SiC(111) are turbostratic, as observed from Raman spectroscopy [21].

The band structure and electronic properties of EG/3C-SiC(111) determined using ARPES showed linear band dispersion K point of the Brillouin zone with the E_F ~500 meV above the E_D indicating n-type monolayer graphene with doping of 10^{13} cm^{-2} [19,66]. The n-type doping is also in agreement with the DFT calculations [21]. Aristov et al. [16] determined the linear band dispersion for EG/3C-SiC(100) using ARPES with the E_F ~250 meV above E_D, which indicated n-type doping. The absence of the buffer layer in EG on 3C-SiC(100) resulted in a weak interaction with the substrate, as evident from E_F closer to E_D compared to EG on 3C-SiC(111) [16].

Transport measurements of EG/3C-SiC were first obtained using field-effect transistor (FET) -based measurements by Moon et al. [80] and Kang et al. using EG formed by thermal decomposition [20]. However, as mentioned in Section 3.3, these measurements are dependent on geometry and electrostatics, and are influenced by the substrate as well as the graphene that has an issue of inconsistent coverage over the substrate [61,102]. In addition, Pradeepkumar et al. reported that the EG grown on 3C-SiC on Si wafers are prone to severe 3C-SiC/Si interface degradation and the lack of large scale continuity of EG over 3C-SiC [106,107]. The prior attempts by Moon et al. and Kang et al. did not address these limitations. The alloy-mediated graphitization using 3C-SiC synthesized on highly-resistive Si wafers from Reference [21] overcame the interface degradation and continuous graphene coverage issues and enabled the large scale transport properties of EG using vdP Hall effect measurements, as given below.

The vdP transport measurements on EG from Reference [21] indicated that the EG-substrate interaction dominates the charge transport within EG. The graphene is strongly p-type doped (carrier concentration ~10^{13} cm^{-2}, mobility ~80 cm^2·V^{-1}·s^{-1} in EG/3C-SiC(100)) as a result of the EG-substrate interaction, which comprises of silicates (charge transfer from EG into the silicates) produced at the interface by the alloy-mediated synthesis. The buffer layer presence in EG/3C-SiC (111) reduces the charge transfer and improve the mobility almost five times compared to EG/3C-SiC (100) at a value of

330 $cm^2 \cdot V^{-1} \cdot s^{-1}$ and carrier concentration of 10^{12} cm^{-2} (p-type). The effect of interface silicates on the charge transport of EG on (100) and (111) 3C-SiC are in agreement with DFT calculations [21].

The transport measurements on EG also indicated that, within the observed diffusion regime (mean free path 3–10 nm), the grain sizes (<100 nm) and the number of graphene layers has no effect on the charge transport of EG. Further support to the lack of correlation of the grain sizes with the EG charge transport characteristics can also be found in the work by Ouerghi et al. [9,10]. A domain size of about 1 µm for EG/3C-SiC(111) was reported for the EG at a sheet carrier concentration of ~2 × 10^{13} cm^{-2} via the linear dispersion of DOS [66]—a value in line with the carrier concentration of EG/3C-SiC(111) from Reference [21] with grain sizes <100 nm.

4.2. Effect of Epitaxial Graphene-Ambient Interaction

The surface of EG is sensitive to a range of gases present in the ambient air [7–12]. Early on, researchers realized that adsorbed atoms or molecules on a graphene device could modify the channel conductivity. Schedin et al. [108] utilized this property to create a detector with sensitivity to very low concentrations of NH_3, NO_2, H_2O_2, and CO_2. Yet, graphene is relatively sensitive to all adsorbed molecules, and, hence, a requirement on building any sensor requires the ability to discriminate the target molecule from the background. An additional implication is that the surface of a graphene sample must be prepared to mitigate the effects of adsorbed molecules on transport properties. For example, heating the sample to temperatures ~400 °C in ultra-high vacuum (UHV) results in a clean, usable surface [109].

In a similar vein, Panchal et al. prepared a reproducible graphene surface to study the effects of adsorbed gases on work function and electrical properties. This led to several important findings. First, there are a range of molecules in the ambient (O_2, H_2O, and NO_2), which act as p-type dopants and reduces the n-type sheet charge density of graphene. Yet, there are other unidentified atmospheric contaminants, which also act as p-type dopants. It was suggested that some candidates are N_2O_4, CO_2, and various hydrocarbons. Lastly, it was shown that, for more than one graphene layer, the effect on conductivity was less, and it was proposed this was due to charge screening of the outermost layer. First-principles modeling indicates that the physical mechanism of the doping compensation (for H_2O, NH_3, CO, NO_2, and NO) and is due to charge transfer [110] from graphene to the molecule when the lowest unoccupied molecular orbital is below the Dirac point.

The polarity of the Si-face SiC gives rise to high hydrophilicity, which is relatively unchanged with the addition of a graphene monolayer [111]. Graphene's response to water and changes in the environment are strongly thickness dependent with a monolayer on Si-face being most sensitive to water adsorption than the bilayer or tri-layer graphene [5,112,113].

Yet, the understanding of the exact H_2O doping mechanism is not as well-understood since chemical processes may be involved. Sidorov et al. noted that, upon ambient exposure at 300 K, both the MEG and MLG on SiC(000$\bar{1}$) typically exhibit p-type conduction, 4.5 × 10^{14} cm^{-2} (Figure 12). The p-type doping is due to the ambient adsorbed film containing water vapour (H_2O), O_2, and NO_2 on the exposed surface [5,114]. The Dirac energy of the graphene lies near the redox potential of dissolved oxygen, which causes the surface charge transfer from graphene into the film (water/oxygen redox couple) and results in compensation of the n-type sheet density [2,5,46,114,115]. In addition, the conductivity state of vacuum annealed EG on SiC(000$\bar{1}$) is n-type (Figure 10b) [112,114]. This n-type conductivity was related to the work function considerations and E_F pinning at the state related to dangling bonds, as discussed in Section 4.1.3 [5]. Yet, we now know that the pseudo charge effect induced by substrate hexagonality dominates the graphene charge state.

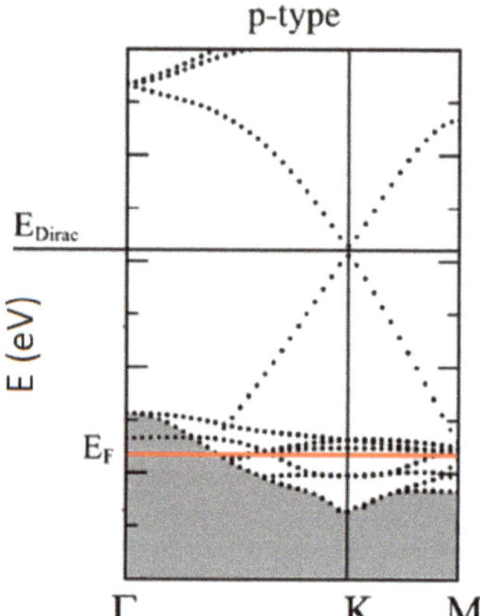

Figure 12. P-type doping of graphene upon the ambient exposure at 300 K. The p-type doping correspond to depletion of 4.5×10^{14} electrons/cm^{-2}. Adapted with permission from Reference [5]. Copyright (2012) AIP Publishing.

Knight et al. exposed monolayer EG on SiC(0001) to different inert gases and performed in-situ, contactless majority free charge carrier determination using terahertz-frequency optical Hall effect measurements [2]. The transport properties of graphene are also sensitive to other adsorbed gas molecules N_2O_4, CO_2, and hydrocarbons [2,5,114,116–118]. The mechanism of ambient doping is described as ambient-exposed graphene forms a thin film of water containing dissolved CO_2 that reacts with water forming H^+ ions. The oxygen in the water film reacts with H^+ ions to form additional water molecules, and electrons are removed from the graphene by p-doping the graphene. This doping is reversible and does not depend on the type of inert gas [2].

4.3. Effect of Epitaxial Graphene-Contact Metal Interaction

Since electronic transport measurements involve making metal contacts, understanding the influence of metal contacts on the transport of EG is also a concern [3]. The work function difference between graphene and a metal induces doping in graphene since the Fermi levels must equilibrate [3].

Since the E_F of freestanding graphene meets the E_D, the adsorption of metal can significantly alter its electronic properties [3], as the DOS of graphene at Dirac point is much lower than that of the metal. A tiny amount of charge transfer shifts the E_F significantly [25,33,34] as it is reported that 0.01 electrons per carbon atom would shift the E_F by 470 meV. The charge transfer creates a dipole layer at the graphene-metal interface with ΔV potential. The value of ΔV depends on the strength of the metal-graphene interaction. Even a weak interaction can cause a large shift in the E_F away from the E_D [3,119].

Khomyakov et al. [119] used DFT to characterize the adsorption of graphene on various metals and found graphene interacts more strongly to Ni, Co, Ti, and Pd (chemisorption). The chemisorption involves hybridization between graphene p_z and metal d-states that cause an opening of a band gap in the graphene. In the case of chemisorption, the band structure is disturbed, and the doping is estimated from the difference of work function between the graphene covered with metal and metal-free graphene. In the case of Ni, Co, Ti, and Pd, the graphene is n-type doped [119]. Physisorption, e.g., interaction with metals such as Ag, Al, Cu, Au, and Pt, involves only weak binding to graphene. Therefore, the electronic structure is preserved. In the case of physisorption, the electrons transfer from the graphene to the metal, which causes the E_F to move closer to the E_D—and the hole doping is estimated from the change. At an equilibrium distance from the metal, the graphene is reported to be p-doped on Au and Pt and n-doped on Al, Ag, and Cu. Khomyakov et al. also reported that the graphene is doped n-type when the metal work function <5.4 eV, while it is p-type when the work function is >5.4 eV [119].

Using femtosecond laser-patterned microstructures (resist free, defect-free method), Nath et al. [120] studied the graphene-metal interactions on EG synthesized on 6H-SiC(0001) via thermal decomposition. They found that the nickel does not form bonds with EG on SiC in contrast to the Ni-C formation in chemical vapour deposition (CVD) grown graphene on metals due to the absence of defect sites [119]. However, it was shown that the contact resistance (R_C) of Ni to EG with resistance residue under the contact is an order of magnitude greater than the intrinsic quantum-limited R_C, which proposes the requirement for both end-contacts as well as a residue-free graphene–metal interface (which can form resistive $NiCO_3$) as requirements to obtain quantum-limited contact resistance [120].

DeJarld et al. [121] calculated the work functions of a number of high purity noble metals such as Cr, Cu, Rh, Ni, and Pt, and rare earth metals such as Pr, Eu, Er, Yb, and Y, in contact with MLG, using Kelvin Probe Force Microscopy. The metals were deposited under manufacturing-like conditions and Au (and its published work function) was used as the standard. The work functions were compared to published values for the other metals measured under careful, i.e., UHV, conditions. For the noble metals, a comparison showed Cr, Cu, and Rh had higher work functions (~0.5 eV) than expected and Ni and Pt were lower than expected (also ~0.5 eV). The low work function rare earth metal set were all affected, to various extents, by oxygen in the ambient and/or reactions to graphene rendering most unusable. The most stable rare earth was Yb, even though protection from the ambient was strongly recommended.

Yang et al. [122] demonstrated that the e-beam evaporated Ni on CVD graphene caused the EG work function to be 0.34 eV below the Ni work function. This resulted in electron transfer from graphene to Ni, which makes the graphene underneath the metal contact p-type.

4.4. Mobility and Sheet Carrier Concentration: Power-Law Relationship

A power-law relationship of mobility and sheet carrier concentration was demonstrated by several research groups for both EG on SiC and EG on 3C-SiC [4,21,25,57,59,123]. Tedesco et al. [56] described the functional power-law relationship between increasing mobility and decreasing carrier density as an intrinsic property of EG. The general power-law behaviour of mobility and sheet carrier concentration indicate that the tunability of graphene transport properties is constrained.

In Reference [21], the authors of this review demonstrate that the mobility versus sheet carrier concentration values for EG/SiC(0001) from Reference [56], and EG/3C-SiC can be fitted with good confidence using the same power law, which demonstrates a common conductivity of ~3 ± 1 (e^2/h) close to the minimum quantum conductivity of graphene (Figure 13). Note that the grain sizes of EG/SiC(0001) are at least 100 nm and more [124–126] whereas the EG/3C-SiC is smaller, which indicates that grain sizes do not determine the transport properties of EG [21].

In addition, Reference [21] demonstrated that the charge transport in EG/3C-SiC is dominated by the EG-substrate interaction resulting in p-type graphene due to a charge transfer from the EG into the interface silicates. The substrate interaction is stronger in EG/3C-SiC (100) and smaller in the case of EG/3C-SiC(111) due to the presence of a buffer layer in between EG and the substrate, which screens the charge transfer up to an extent. Figure 13 also indicates the different power-law nature of C-face EG from that of the Si-face EG and EG on 3C-SiC, which is due to distinct levels of EG-substrate interaction, according to Norimatsu et al. [11]. This was confirmed by Pradeepkumar et al. [21].

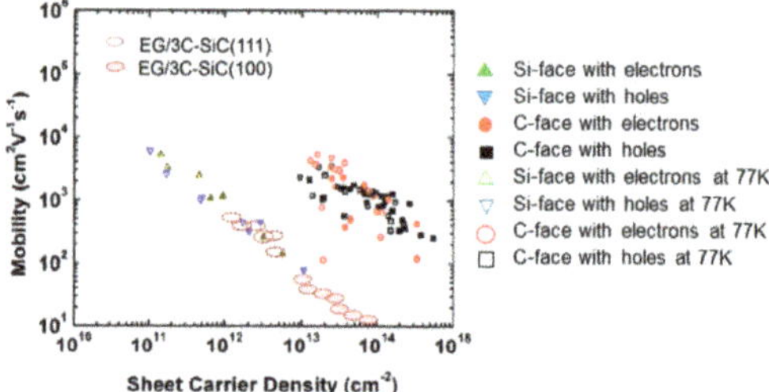

Figure 13. Mobility and sheet carrier density for EG on 3C-SiC/Si combined with those of monolayer EG on the Si-face of 4H-SiC and 6H-SiC substrates at 300 and 77 K from Tedesco et al. [36]. Reprinted with permission from Reference [21]. Copyright (2020) American Chemical Society.

4.5. Temperature-Dependent Transport: Scattering Mechanisms

Temperature-dependent Hall effect measurements in EG/SiC and 3C-SiC are crucial to investigate the charge transport as it provides information on the scattering mechanisms. This section summarizes the temperature-dependent transport properties of EG on Si-face SiC, EG on C-face SiC, and EG on 3C-SiC.

In general, Hwang et al. [93] reported that the carrier scattering mechanism dominant in MLG (on SiO_2) is Coulomb scattering by randomly charged impurities at the graphene-substrate interface, as also reported by Chen et al. [127]. The same group determined the typical random charged impurity concentration to be ~10^{12} cm^2 and suggested that, by reducing this impurity concentration to ~10^{10} cm^2 range, it should lead to a significant increase in the EG mobility. The charge transport in graphene dominated by the charged impurities in the substrate can be theoretically described by the Drude-Boltzmann model [93], which is valid only in the high-density regime [128]. The Fermi temperature of graphene is about 1300 K for n~10^{12} cm^2, which indicates no temperature dependence of the MLG conductivity arising from charged impurity scattering within a 0–300 K temperature range in agreement with the experimental observation of graphene flakes [93].

In contrast to the graphene flake results, MLG on Si-face SiC shows a strong temperature dependence of mobility and resistivity, as shown in Figure 14 [83,123,129]. At carrier concentrations away from the charge neutrality point (above 10^{11} cm^{-2}), the mobility decreases with increasing the temperature and the resistivity increases [123]. A remote phonon scattering originating from the SiC substrate including the buffer layer and the C-Si bonds at the interface was reported as the reason for the strong temperature dependence of mobility and resistivity, and the limitation of both properties at high temperatures [26,123,130,131]. Tedesco et al. [56] reported that the suppression of conductivity in high temperatures in the case of monolayer or bilayer Si-face EG is due to the scattering from the point defects present in the interfacial layer.

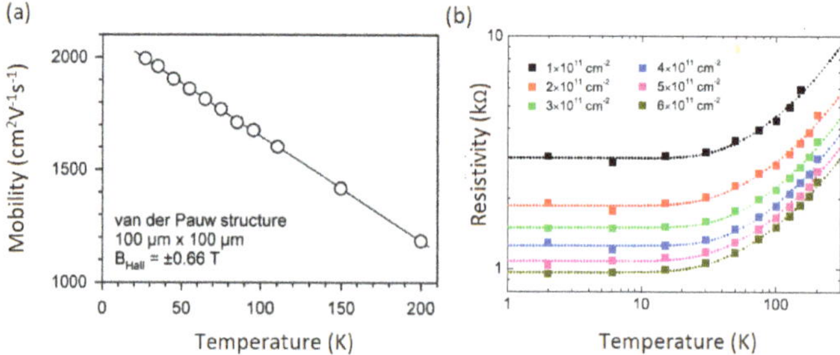

Figure 14. (a) Temperature dependent mobility of MLG on SiC (0001). Reprinted with permission from Reference [26]. Copyright (2009) Springer Nature. (b) Temperature dependent resistivity of MLG on SiC (0001). Reprinted with permission from Reference [123].

Figure 15 indicates the sheet resistance of rotationally disordered MEG (~10 layers) formed on SiC(000$\bar{1}$) [132], which indicates that the sheet resistance remains nearly stable within 0–300 K. This has been related to the weak substrate interaction of EG on the C-face SiC [11]. The mobility, in this case, does not significantly depend on temperature [8]. The phonon scattering is suppressed, and the scattering from impurities is weak.

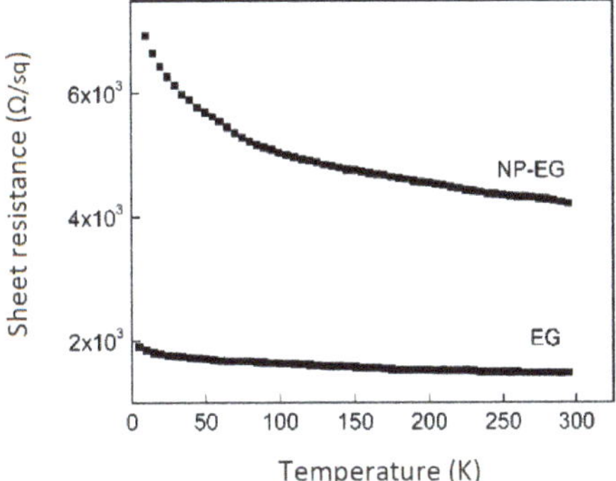

Figure 15. Temperature dependent sheet resistance of EG on SiC(000$\bar{1}$) and nitrophenyl functionalized graphene (NP-EG). Reprinted with permission from Reference [132]. Copyright (2009) American Chemical Society.

Figure 16 shows mobility versus temperature for EG/3C-SiC substrates [21]. The mobility of EG/3C-SiC (100) shows only a weak temperature dependence, slowly increasing after 200 K and reaching a value of ~30 cm$^2 \cdot$V$^{-1} \cdot$s^{-1} at 300 K. This indicates charge impurity scattering. The weak temperature dependence of the mobility of EG/3C-SiC(100) is consistent with Reference [83] for EG on SiC, and is attributed to the absence of the buffer layer. In the case of EG/3C-SiC(111), the mobility shows a sharp increase after 200 K up to a value of ~375 cm$^2 \cdot$V$^{-1} \cdot$s^{-1} at 250 K, which, again, indicates charge impurity scattering, and then a decrease to ~330 cm$^2 \cdot$V$^{-1} \cdot$s^{-1} at 300 K. The negative temperature dependence of mobility for the EG/3C-SiC(111) above 250 K has been related to scattering at the buffer layer in EG on 6H-SiC [133]. At low temperatures, the charge impurity scattering is dominated [56,127] and, since the temperature is increased, the scattering decreases [25]. In addition, the constant value of conductivity (~3 ± 1 (e^2/h)) obtained for EG/3C-SiC also indicates that the charge impurity scattering at the interface dominates the transport [21].

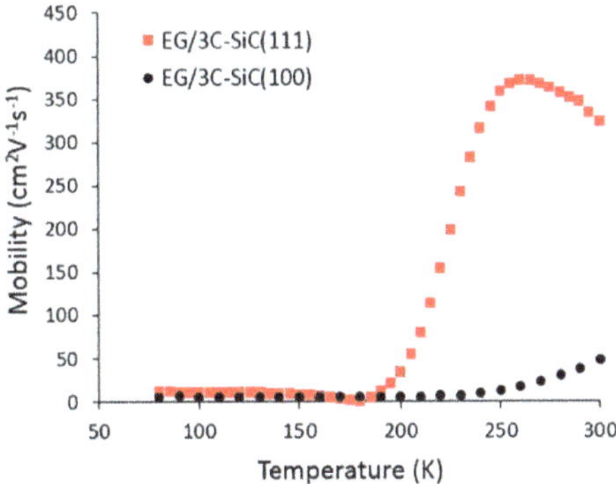

Figure 16. Temperature dependent mobility between 80 and 300 K for (100) and (111) EG/3C-SiC. Reprinted from Reference [21]. Copyright (2020) American Chemical Society.

5. Fine-Tuning of Transport Properties

Epitaxial graphene on SiC and 3C-SiC suffers from strong intrinsic n-type doping. The substrate-induced n-type doping in EG/SiC and EG/3C-SiC translates into the shift in the E_F, away from the E_D, such that the ambipolar properties of graphene cannot be exploited [79]. The substrate interaction causes larger charge density and lower mobility with strong temperature dependence when compared to freestanding graphene. Some of these issues can be compensated by tuning of the transport properties using different methods discussed in the following section.

5.1. Intercalation

Intercalation has been performed via different elements such as hydrogen [21,75,77,81,134], oxygen [78], fluorine [135], gold [136], and magnesium [43]. The monolayer graphene formed after intercalation of buffer layer on SiC (0001) is called quasi-free-standing monolayer graphene (QFMLG). Riedl et al. [134] produced QFMLG by annealing the buffer layer on Si-face SiC(0001) in atmospheric pressure H_2. The XPS C 1s and Si 2p spectra demonstrated that H saturate the Si dangling bonds and decouples the buffer layer by forming a monolayer graphene [134]. Buffer layer decoupling in EG on 3C-SiC (111) via H-intercalation has also been demonstrated [21,76].

The QFMLG on SiC (0001) films measured by ARPES are generally p-type doped [75] (see Figure 17). As discussed in Section 4.1.1, Mammadov et al. and Ristein et al. determined that the p-type conduction of QFMLG is caused by the spontaneous polarization of the SiC [75,81]. QFMLG layers formed on semi-insulating 4H-SiC (0001) indicated a hole concentration of 8.6×10^{12} cm^{-2} (stronger spontaneous polarization of 4H-SiC) extracted from the Fermi surface measurement [75] at E_F 340 meV below the E_D, whereas the QFMLG on 6H-SiC (0001) exhibits a p-type carrier concentration of 5.5×10^{12} cm^{-2} with E_F 300 meV below the E_D. The Fermi velocity was measured to be 0.98×10^6 ms^{-1} [33,75].

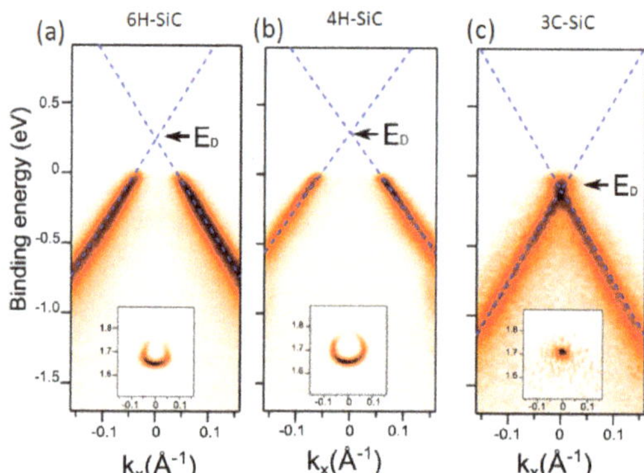

Figure 17. ARPES band structure of QFMLG formed on (**a**) 6H SiC, (**b**) 4H-SiC, and (**c**) 3C-SiC. Reprinted with permission from Reference [75]. Copyright (2014) IOP Publishing Ltd.

Speck et al. demonstrated that the mobility of QFMLG showed reduced temperature dependence compared with an as-grown MLG, which indicates reduced scattering from the underlying SiC (see Figure 18). The mobility of MLG falls by more than 50% when the temperature is increased from 25 K to 300 K whereas the change in mobility is only ~ 10% for the QFMLG across the same temperature range (samples indicated constant carrier concentration across the entire temperature range). The weak temperature dependence of mobility, approaching the behavior exhibited by flakes (see Section 4.5) signifies QFMLG is effectively decoupled from substrate compared to MLG [129]. Furthermore, for T greater than 290 K, the mobility of QFMLG are generally 3–5 times larger than EG before H-intercalation (Figure 18). The production of quasi-free-standing bilayer graphene by H-intercalation of MLG has also been achieved [77,137]. Reference [77] reported H-intercalated BLG with a p-type carrier concentration of 10^{12} cm^{-2} with E_F 230 meV below the E_D. The Fermi velocity was measured to be 1.04×10^6 ms^{-1}.

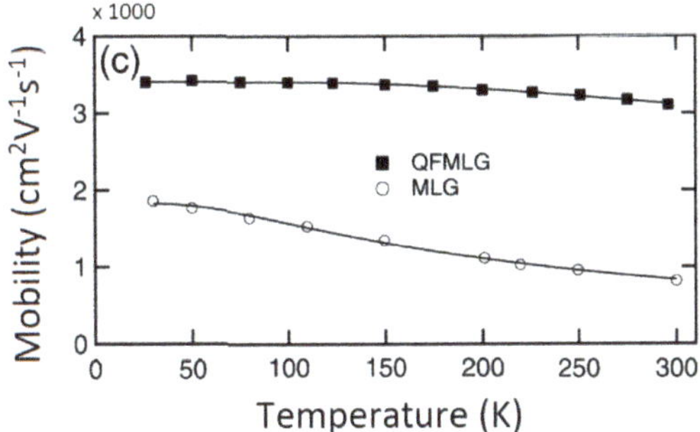

Figure 18. Temperature dependent mobility of QFMLG ($p = 5.7 \times 10^{12}$ cm^{-2}) and MLG (10^{13} cm^{-2}). Reprinted with permission from Reference [129]. Copyright (2011) AIP Publishing.

In contrast to the p-type polarization doping of QFMLG on bulk SiC substrates, the ARPES data of QFMLG on 3C-SiC (111) by H-intercalation reported slight n-type doping (see Figure 17c). This is due to the absence of spontaneous polarization in 3C-SiC [75]. The n-type doping density of about 10^{12} cm^{-2} was estimated, which was related to the residual defects at the interface [76]. Pradeepkumar et al. reported that the p-type transport properties of EG on 3C-SiC (111) did not improve after decoupling the buffer layer, which confirms that the interface silicates dominate the transport properties in this instance.

The intercalation process has also attempted via an oxidation process for EG on SiC (0001). Mathieu et al. [78] demonstrated that, through an oxidation process, the oxygen saturates Si dangling bonds and breaks some Si–C bonds of the buffer layer. In this case, the buffer layer is only partially decoupled, and the resulting sample had molecular oxygen intercalated between carbon layers with a ($\sqrt{3} \times \sqrt{3}$) rotated 30° pattern as observed with μLEED. The same group also demonstrated the possibility to tune the charge density modulation in the graphene by controlling the oxidation parameters. The starting sample had an n-type sheet charge of 1.4×10^{13} cm^{-2} corresponding to E_F of 500 meV below E_D. After the first O-intercalation step, the sheet density changed to 4.6×10^{12} cm^{-2} (n-type) and E_F of 300 meV. The second O-intercalation step yielded 1.9×10^{12} cm^{-2} (n-type) and E_F of 200 meV. Quasi-free-standing bilayer graphene (QFSBLG) with p-type doping of 10^{13} cm^{-2} was achieved by Oliveira Jr et al. via oxygen intercalation by annealing the monolayer graphene on SiC(0001) at 600 °C in the air [138].

The most advanced intercalation may be magnesium intercalation of monolayer EG on SiC (0001) creating QFSBLG [43]. For this case, ARPES measurements demonstrate a very high n-type concentration of 2×10^{14} cm^{-2} at an E_F of 720 meV above E_D and the creation of a band gap of 0.36 eV. The Fermi velocity was measured to be 0.97×10^6 ms^{-1}, which is only slightly different from MLG. This may indicate that the graphene lattice is only slightly perturbed by the magnesium intercalant.

Wong et al. [135] obtained a charge-neutral QFSG on SiC (0001) using fluorine from a fluorinated fullerene source, which was stable under ambient conditions, whereas resistant to temperatures up to 1200 °C. Gierz et al. [136] demonstrated buffer-layer decoupling using gold intercalation of EG on SiC (0001). The formation of an n-doped or a p-doped graphene layer can be controlled by varying the gold coverage about one-third or one monolayer, respectively.

5.2. Functionalization of EG

Controlling the transport properties via surface functionalization makes use of charge compensation from surface adsorption or attachments [139]. Coletti et al. demonstrated that the intrinsic n-type doping of EG on SiC (0001) can be compensated by functionalizing graphene (non-covalent) with an electron acceptor called as tetrafluorotetracyanoquinodimethane (F4-TCNQ) [79]. Surface transfer doping results in charge transfer from graphene into the doping molecule reducing the sheet density. A 0.8-nm-thick layer of F4-TCNQ on monolayer EG resulted in near charge neutrality, i.e., $E_F = E_D$, Figure 19. For a bilayer sample, the band gap more than doubled, moving from about 100 meV to 250 meV [79] with the E_F shifted into the band gap. Yet, both the compensation and band gap widening saturated near 0.8 nm, which indicated that there may be limitations with this approach. An increase of mobility with decreasing carrier density was seen in the F4-TCNQ doped samples by Jobst et al. [55]. When the carrier density is decreased, the mobility limited by the electron interaction with substrate phonons rises substantially (up to 29,000 cm^2·V^{-1}·s^{-1} at 25 K), and Shubnikov–de Haas oscillations and the graphene-like quantum Hall effect are observed [55]. The F4-TCNQ experiments demonstrate that the effects of the substrate interaction can be mitigated by reducing the sheet density, and the quantum Hall effect and pseudo-relativistic physics and linear band structure are possible in the epitaxial graphene.

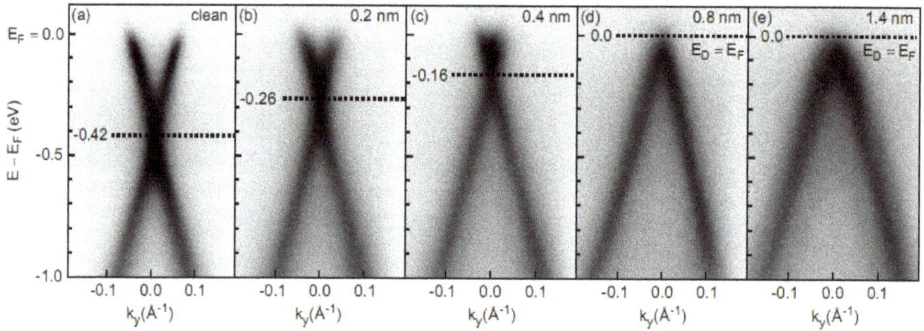

Figure 19. ARPES measured π bands for around K point for the Brillouin zone for (**a**) as-grown graphene monolayer on SiC (0001) (**b–e**) increasing amount of F4-TCNQ. Reprinted with permission from Reference [79].

Bekyarova et al. [132] studied the transport properties of nitrophenylfunctionalized EG (NP-EG formed on the C-face SiC) and suggested that it changes the graphene from near-metallic to semiconducting (see Figure 15) [132].

Other interesting work by Lartsev et al. [140] demonstrate charge neutrality (10^{11} cm^{-2}) and p-doping (5×10^{12} cm^{-2}) by reversible tuning of carrier density via high electrostatic potential gating with ions produced by corona discharge in initially strongly n-doped EG/SiC devices. In this case, the ions deposited on the dielectric layer induce a surface charge density on the graphene. The corona effect is reversible and, depending on the ionic charge, both p-doping and n-doping are possible. The method is utilized in graphene applications such as quantum resistance metrology where specific fixed doping is required.

Peles-Lemli et al. [141] used DFT calculations to investigate the graphene interactions with the adsorbed alkali metal ions such as Li$^+$, Na$^+$, and K$^+$ for the applications of graphene-based chemical sensors. In the presence of an external electric field, the positively charged ions get closer to the graphene inducing excess charge transfer from the graphene surface to the metal ions, causing negative charges on the ions and removal of the ions from the surface. Ludbrook et al. [142] demonstrated first direct observation of superconductivity (critical temperature of 5.9 K) for monolayer graphene formed on 6H-SiC (0001) decorated with a layer of lithium atoms at a low temperature, which caused an enhanced electron-phonon coupling.

5.3. Top-Gate Graphene Field-Effect Transistors

The ability to control the graphene transport properties via an external voltage is the heart of future electronics [29]. Graphene field-effect transistors show ambipolar field-effect due to its unique electronic band structure with a zero band gap [58]. Under a gate voltage control, charge carriers in graphene can be tuned continuously between electrons and holes with the Fermi level crossing the Dirac point [143]. Charge carriers can be tuned up to 10^{13} cm^{-2} with mobilities over 15,000 cm$^2 \cdot$V$^{-1} \cdot$s^{-1} at 300 K [29]. The mobilities depend on temperature weakly, which indicates charge impurity scattering and, therefore, the gate control may be able to improve those numbers further, even up to $\approx$100,000 cm$^2 \cdot$V$^{-1} \cdot$s^{-1} [58].

The first report of epitaxial graphene field-effect transistors (EG FETs) patterned on a single epitaxial graphene chip both on the SiC (0001) and SiC (000$\overline{1}$) was made in 2008 [60]. Figure 20 demonstrates the Id-Vg characteristics of the C-face and Si-face transistors. The results from this work show that the gate-controlled tuning can lead to mobility as high as 5000 cm$^2 \cdot$V$^{-1} \cdot$s^{-1} on the C-face, and an on/off ratio up to 7 on the Si-face SiC. Moon et al. [144] measured the first small-signal radio-frequency (RF) performance of EG FETs on 6H-SiC(0001). Lin et al. [145] fabricated high-speed graphene RF FETs on 4H-SiC (0001) and obtained a cutoff frequency of 100 GHz for a gate length of 240 nm. He et al. [146] revealed the RF performance of bilayer GFETs up to 200 °C. Yu et al. [147] formed a QFSBLG FET on 4H-SiC (0001), (100-nm gate length) with an intrinsic current gain cutoff frequency 407 GHz. Hwang et al. [148] demonstrated the first top-gated graphene nanoribbon FETs of 10 nm width on 6H-SiC (0001). The results exhibited exceptionally high drive currents (for fast switching) and opening of a substantial band gap of 0.14 eV. Bianco et al. [149] determined the detection of terahertz radiation at room temperature using antenna-coupled EG FETs on SiC, which introduced the potential of plasmonic detectors using epitaxial graphene on silicon carbide.

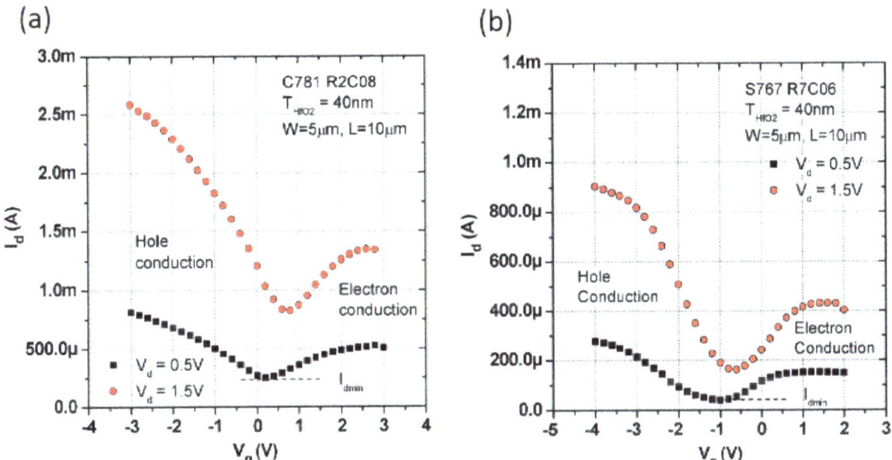

Figure 20. Id$-$Vg characteristics for FET fabricated on (**a**) MEG on SiC (000$\overline{1}$) transistor for 0.5 V and 1.5 V drain voltages, (**b**) EG on SiC (0001) transistor for two drain voltages. Reprinted with permission from Reference [60]. Copyright (2008) Institute of Electrical and Electronics Engineers.

Additionally, the gate control in FET fabricated on bilayer EG enables us to selectively adjust the carrier concentration in each layer and induce a band gap in bilayer graphene. Ohta et al. demonstrated that, by doping the bilayer of graphene, the band gap of 200 meV could be achieved [34]. Alternatively, Zhang et al. showed the band gap of up to 250 meV is achievable by varying the electric field and by using an external gate control. Zhou et al. [45] reported a substrate-induced energy gap of 260 meV [37].

Moon et al. demonstrated the first ambipolar behavior of wafer-scale EG FET on 3C-SiC(111)/Si substrates formed via thermal decomposition (see Figure 21) [80]. A comparison of the graphene FETs fabricated on Si(110) and Si(111) substrates were reported by Kang et al. [20]. The graphene FETs on Si(111) exhibited higher channel currents, which is an order of magnitude larger than those on Si(110) substrates. However, the measurements indicated the presence of a large amount of gate-leakage current.

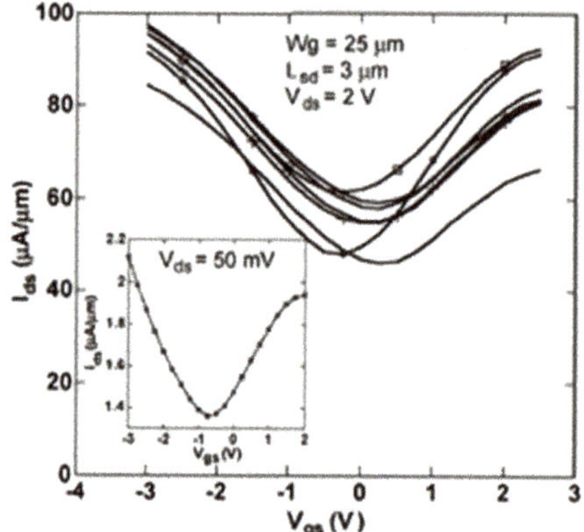

Figure 21. Ambipolar field-effect transistor on Si (111) substrates. Reprinted with permission from Reference [80]. Copyright (2010) Institute of Electrical and Electronics Engineers.

Fabrication of graphene FETs involves the challenges of selecting an appropriate gate insulator and controlling its thickness [150]. Generally, the oxides such as SiO_2, HfO_2, Al_2O_3, and Y_2O_3 are used as the gate dielectrics, but these tend to become amorphous with scaling and interactions with the graphene can occur, which makes the high-quality defect-free interface formation difficult [150–152]. Hexagonal boron nitride (h-BN) forms a weak van der Waals interaction with the graphene (lattice mismatch only 2% [153]) and has been utilized as an ideal insulator for the graphene FET formation [151]. Due to the lower value of a dielectric constant (k~4), h-BN is usually combined with a high k-dielectric layer such as Y_2O_3 formed via atomic layer deposition on h-BN [154]. The scaling (~1.3 nm) of h-BN results in excessive gate leakage current within the FET. To overcome this, Illarionov et al. [150] utilized epitaxial calcium fluoride of ~2 nm thickness as a gate insulator with low leakage current and on/off ratio up to 10^7. Furthermore, the epitaxial growth of graphene-h-BN heterostructures on copper via chemical vapor deposition has been demonstrated by Geng et al. [155].

6. Conclusions

Epitaxial graphene is a practical approach for electronics and photonic applications thanks to its wafer-scale integration compatibility [5]. The knowledge and the control of electronic and transport properties of epitaxial graphene are crucial for graphene-based electronic and photonics applications.

This work has reviewed the impact of different interactions the epitaxial graphene makes with its surroundings, such as substrate, ambient, and metal contacts. The review concluded that the electronic and charge transport properties of EG on both SiC and Si substrates are always dominated by the interfaces between graphene and its environment. Within the diffusion regime, the grain sizes do not affect carrier transport in EG. The sheet carrier concentration and the mobility of graphene follow a power-law dependence, which is universal for the substrate-supported graphene. Lastly, different methods to tune the electronic and transport properties of EG are discussed, which points out that the control of the charge transfer/doping in EG is directly linked for engineering the graphene interfaces. Hence, it is critical for the integration of graphene into future micro-electronic or nano-electronic devices.

Author Contributions: The manuscript was prepared through the contributions from A.P., D.K.G., and F.I. All authors have read and agreed to the published version of the manuscript.

Funding: A.P. and F.I. kindly acknowledge funding from the Air Force Office for Scientific Research through the Asian Office for Aerospace Research and Development (AOARD, grant 18IOA052).

Acknowledgments: The authors gratefully acknowledge discussions with Antonija Grubišić-Čabo, Jimmy Kotsakidis, and Michael Fuhrer from the Centre of Excellence on Future Low-Energy Electronics Technologies, Monash University, Melbourne, Australia.

Conflicts of Interest: The authors declare no conflict of interest.

References

1. Motta, N.; Iacopi, F.; Coletti, C. *Growing Graphene on Semiconductors*; Pan Stanford: Singapore, 2017.
2. Knight, S.; Hofmann, T.; Bouhafs, C.; Armakavicius, N.; Kühne, P.; Stanishev, V.; Ivanov, I.G.; Yakimova, R.; Wimer, S.; Schubert, M.; et al. In-situ terahertz optical Hall effect measurements of ambient effects on free charge carrier properties of epitaxial graphene. *Sci. Rep.* **2017**, *7*, 1–8. [CrossRef]
3. Giovannetti, G.; Khomyakov, P.; Brocks, G.; Karpan, V.M.; Van den Brink, J.; Kelly, P.J. Doping graphene with metal contacts. *Phys. Rev. Lett.* **2008**, *101*, 026803. [CrossRef]
4. Berger, C.; Song, Z.; Li, T.; Li, X.; Ogbazghi, A.Y.; Feng, R.; Dai, Z.; Marchenkov, A.N.; Conrad, E.H.; First, P.N.; et al. Ultrathin epitaxial graphite: 2D electron gas properties and a route toward graphene-based nanoelectronics. *J. Phys. Chem. B* **2004**, *108*, 19912–19916. [CrossRef]
5. Sidorov, A.N.; Gaskill, K.; Buongiorno Nardelli, M.; Tedesco, J.L.; Myers-Ward, R.L.; Eddy, C.R., Jr.; Jayasekera, T.; Kim, K.W.; Jayasingha, R.; Sherehiy, A.; et al. Charge transfer equilibria in ambient-exposed epitaxial graphene on (0001⁻) 6H-SiC. *J. Appl. Phys.* **2012**, *111*, 113706. [CrossRef]
6. Nishi, Y. Semiconductor Interfaces and Their Implications to VLSI Device Reliability. In *Control of Semiconductor Interfaces*; Elsevier: Amsterdam, The Netherlands, 1994; pp. 9–12.
7. Rhoderick, E. Surfaces and Interfaces in Semiconductor Technology. In *Physics and Contemporary Needs*; Springer: Berlin/Heidelberg, Germany, 1978; pp. 3–28.
8. Berger, C.; Song, Z.; Li, X.; Wu, X.; Brown, N.; Naud, C.; Mayou, D.; Li, T.; Hass, J.; Marchenkov, A.N.; et al. Electronic confinement and coherence in patterned epitaxial graphene. *Science* **2006**, *312*, 1191–1196. [CrossRef] [PubMed]
9. Hannon, J.; Copel, M.; Tromp, R. Direct Measurement of the Growth Mode of Graphene on SiC (0001) and SiC (000 1). *Phys. Rev. Lett.* **2011**, *107*, 166101. [CrossRef] [PubMed]
10. Seyller, T.; Bostwick, A.; Emtsev, K.; Horn, K.; Ley, L.; McChesney, J.; Ohta, T.; Riley, J.D.; Rotenberg, E.; Speck, F. Epitaxial graphene: A new material. *Phys. Status Solidi B* **2008**, *245*, 1436–1446. [CrossRef]
11. Norimatsu, W.; Kusunoki, M. Structural features of epitaxial graphene on SiC {0 0 0 1} surfaces. *J. Phys. D Appl. Phys.* **2014**, *47*, 094017. [CrossRef]
12. Hass, J.; Varchon, F.; Millán-Otoya, J.; Sprinkle, M.; de Heer, W.; Berger, C.; First, P.; Magaud, L.; Conrad, E. Rotational stacking and its electronic effects on graphene films grown on 4H-SiC (0001̄). *arXiv* **2007**, arXiv:0706.2134.
13. Hite, J.K.; Twigg, M.E.; Tedesco, J.L.; Friedman, A.L.; Myers-Ward, R.L.; Eddy, C.R., Jr.; Gaskill, D.K. Epitaxial graphene nucleation on C-face silicon carbide. *Nano Lett.* **2011**, *11*, 1190–1194. [CrossRef] [PubMed]
14. Lin, Y.-M.; Dimitrakopoulos, C.; Farmer, D.B.; Han, S.-J.; Wu, Y.; Zhu, W.; Gaskill, D.K.; Tedesco, J.L.; Myers-Ward, R.L.; Eddy, C.R., Jr.; et al. Multicarrier transport in epitaxial multilayer graphene. *Appl. Phys. Lett.* **2010**, *97*, 112107. [CrossRef]
15. Wu, X.; Hu, Y.; Ruan, M.; Madiomanana, N.K.; Hankinson, J.; Sprinkle, M.; Berger, C.; de Heer, W.A. Half integer quantum Hall effect in high mobility single layer epitaxial graphene. *Appl. Phys. Lett.* **2009**, *95*, 223108. [CrossRef]
16. Aristov, V.Y.; Urbanik, G.; Kummer, K.; Vyalikh, D.V.; Molodtsova, O.V.; Preobrajenski, A.B.; Zakharov, A.A.; Hess, C.; Hänke, T.; Buchner, B.; et al. Graphene synthesis on cubic SiC/Si wafers. Perspectives for mass production of graphene-based electronic devices. *Nano Lett.* **2010**, *10*, 992–995. [CrossRef] [PubMed]
17. Suemitsu, M.; Fukidome, H. Epitaxial graphene on silicon substrates. *J. Phys. D Appl. Phys.* **2010**, *43*, 374012. [CrossRef]
18. Gupta, B.; Notarianni, M.; Mishra, N.; Shafiei, M.; Iacopi, F.; Motta, N. Evolution of epitaxial graphene layers on 3C SiC/Si (1 1 1) as a function of annealing temperature in UHV. *Carbon* **2014**, *68*, 563–572. [CrossRef]
19. Ouerghi, A.; Marangolo, M.; Belkhou, R.; El Moussaoui, S.; Silly, M.; Eddrief, M.; Largeau, L.; Portail, M.; Fain, B.; Sirotti, F. Epitaxial graphene on 3C-SiC (111) pseudosubstrate: Structural and electronic properties. *Phys. Rev. B* **2010**, *82*, 125445. [CrossRef]

20. Kang, H.-C.; Karasawa, H.; Miyamoto, Y.; Handa, H.; Fukidome, H.; Suemitsu, T.; Suemitsu, M.; Otsuji, T. Epitaxial graphene top-gate FETs on silicon substrates. *Solid State Electron.* **2010**, *54*, 1071–1075. [CrossRef]

21. Pradeepkumar, A.; Amjadipour, M.; Mishra, N.; Liu, C.; Fuhrer, M.S.; Bendavid, A.; Isa, F.; Zielinski, M.; Sirikumara, H.I.; Jayasekera, T.; et al. p-type Epitaxial Graphene on Cubic Silicon Carbide on Silicon for Integrated Silicon Technologies. *ACS Appl. Nano Mater.* **2019**. [CrossRef]

22. Wallace, P.R. The band theory of graphite. *Phys. Rev.* **1947**, *71*, 622. [CrossRef]

23. Hass, J.; Varchon, F.; Millan-Otoya, J.-E.; Sprinkle, M.; Sharma, N.; de Heer, W.A.; Berger, C.; First, P.N.; Magaud, L.; Conrad, E.H. Why multilayer graphene on 4 H– SiC (000 1) behaves like a single sheet of graphene. *Phys. Rev. Lett.* **2008**, *100*, 125504. [CrossRef] [PubMed]

24. Neto, A.H.C.; Guinea, F.; Peres, N.M.R.; Novoselov, K.S.; Geim, A.K. The electronic properties of graphene. *Rev. Mod. Phys.* **2009**, *81*, 109. [CrossRef]

25. Zhu, W.; Perebeinos, V.; Freitag, M.; Avouris, P. Carrier scattering, mobilities, and electrostatic potential in monolayer, bilayer, and trilayer graphene. *Phys. Rev. B* **2009**, *80*, 235402. [CrossRef]

26. Emtsev, K.V.; Bostwick, A.; Horn, K.; Jobst, J.; Kellogg, G.L.; Ley, L.; McChesney, J.L.; Ohta, T.; Reshanov, S.A.; Röhrl, J.; et al. Towards wafer-size graphene layers by atmospheric pressure graphitization of silicon carbide. *Nat. Mater.* **2009**, *8*, 203–207. [CrossRef] [PubMed]

27. Nakada, K.; Fujita, M.; Dresselhaus, G.; Dresselhaus, M.S. Edge state in graphene ribbons: Nanometer size effect and edge shape dependence. *Phys. Rev. B* **1996**, *54*, 17954. [CrossRef] [PubMed]

28. Zhang, Y.; Tan, Y.-W.; Stormer, H.L.; Kim, P. Experimental observation of the quantum Hall effect and Berry's phase in graphene. *Nature* **2005**, *438*, 201–204. [CrossRef]

29. Novoselov, K.S.; Geim, A.K.; Morozov, S.V.; Jiang, D.; Zhang, Y.; Dubonos, S.V.; Grigorieva, I.V.; Firsov, A.A. Electric field effect in atomically thin carbon films. *Science* **2004**, *306*, 666–669. [CrossRef] [PubMed]

30. Mattausch, A.; Pankratov, O. Ab initio study of graphene on SiC. *Phys. Rev. Lett.* **2007**, *99*, 076802. [CrossRef]

31. Varchon, F.; Feng, R.; Hass, J.; Li, X.; Nguyen, B.N.; Naud, C.; Mallet, P.; Veuillen, J.-Y.; Berger, C.; Conrad, E.H.; et al. Electronic structure of epitaxial graphene layers on SiC: Effect of the substrate. *Phys. Rev. Lett.* **2007**, *99*, 126805. [CrossRef]

32. Bostwick, A.; Ohta, T.; Seyller, T.; Horn, K.; Rotenberg, E. Quasiparticle dynamics in graphene. *Nat. Phys.* **2007**, *3*, 36–40. [CrossRef]

33. Ohta, T.; Bostwick, A.; McChesney, J.L.; Seyller, T.; Horn, K.; Rotenberg, E. Interlayer interaction and electronic screening in multilayer graphene investigated with angle-resolved photoemission spectroscopy. *Phys. Rev. Lett.* **2007**, *98*, 206802. [CrossRef]

34. Ohta, T.; Bostwick, A.; Seyller, T.; Horn, K.; Rotenberg, E. Controlling the electronic structure of bilayer graphene. *Science* **2006**, *313*, 951–954. [CrossRef]

35. Sprinkle, S.M.; Siegel, D.; Hu, Y.; Hicks, J.; Tejeda, A.; Taleb-Ibrahimi, A.; le Fèvre, P.; Bertran, F.; Vizzini, S.; Enriquez, H.; et al. First direct observation of a nearly ideal graphene band structure. *Phys. Rev. Lett.* **2009**, *103*, 226803. [CrossRef] [PubMed]

36. Starke, U.; Riedl, C. Epitaxial graphene on SiC (0001) and: From surface reconstructions to carbon electronics. *J. Phys. Condens. Matter.* **2009**, *21*, 134016. [CrossRef]

37. Zhou, S.Y.; Gweon, G.-H.; Fedorov, A.; de First, P.; de Heer, W.; Lee, D.-H.; Guinea, F.; Neto, A.C.; Lanzara, A. Substrate-induced bandgap opening in epitaxial graphene. *Nat. Mater.* **2007**, *6*, 770–775. [CrossRef] [PubMed]

38. Bychkov, Y.A.; Martinez, G. Magnetoplasmon excitations in graphene for filling factors $v \leq 6$. *Phys. Rev. B* **2008**, *77*, 125417. [CrossRef]

39. Davydov, S.Y. Estimations of the Fermi Velocity and Effective Mass in Epitaxial Graphene and Carbyne. *Tech. Phys. Lett.* **2019**, *45*, 650–652. [CrossRef]

40. McCann, E.; Fal'ko, V.I. Landau-level degeneracy and quantum Hall effect in a graphite bilayer. *Phys. Rev. Lett.* **2006**, *96*, 086805. [CrossRef]

41. Sarma, S.D.; Hwang, E.; Rossi, E. Theory of carrier transport in bilayer graphene. *Phys. Rev. B* **2010**, *81*, 161407. [CrossRef]

42. McCann, E.; Koshino, M. The electronic properties of bilayer graphene. *Rep. Prog. Phys.* **2013**, *76*, 056503. [CrossRef]

43. Grubišić-Čabo, A.; Kotsakidis, J.C.; Yin, Y.; Tadich, A.; Haldon, M.; Solari, S.; di Bernardo, I.; Daniels, K.M.; Riley, J.; Huwald, E.; et al. Magnesium-intercalated graphene on SiC: Highly n-doped air-stable bilayer graphene at extreme displacement fields. *Mesoscale Nanoscale Phys.* **2020**, arXiv:2005.02670.

44. Novoselov, K.S.; McCann, E.; Morozov, S.; Fal'ko, V.I.; Katsnelson, M.; Zeitler, U.; Jiang, D.; Schedin, F.; Geim, A. Unconventional quantum Hall effect and Berry's phase of 2π in bilayer graphene. *Nature Phys.* **2006**, *2*, 177–180. [CrossRef]

45. Zhang, Y.; Tang, T.-T.; Girit, C.; Hao, Z.; Martin, M.C.; Zettl, A.; Crommie, M.F.; Shen, Y.R.; Wang, F. Direct observation of a widely tunable bandgap in bilayer graphene. *Nature* **2009**, *459*, 820–823. [CrossRef]

46. Escobedo-Cousin, E.; Vassilevski, K.; Hopf, T.; Wright, N.; O'Neill, A.; Horsfall, A.; Goss, J.; Cumpson, P. Local solid phase growth of few-layer graphene on silicon carbide from nickel silicide supersaturated with carbon. *J. Appl. Phys.* **2013**, *113*, 114309. [CrossRef]

47. Jayasekera, T.; Kim, K.; Nardelli, M.B. Electronic and Structural Properties of Turbostratic Epitaxial Graphene on the 6H-SiC (000-1) Surface. *Mater. Sci. Forum* **2012**, *717–720*, 595–600. [CrossRef]

48. Siegel, D.A.; Hwang, C.; Fedorov, A.V.; Lanzara, A. Quasifreestanding multilayer graphene films on the carbon face of SiC. *Phys. Rev. B* **2010**, *81*, 241417. [CrossRef]

49. Orlita, M.; Faugeras, C.; Plochocka, P.; Neugebauer, P.; Martinez, G.; Maude, D.K.; Barra, A.-L.; Sprinkle, M.; Berger, C.; de Heer, W.A.; et al. Approaching the Dirac point in high-mobility multilayer epitaxial graphene. *Phys. Rev. Lett.* **2008**, *101*, 267601. [CrossRef]

50. Miller, D.L.; Kubista, K.D.; Rutter, G.M.; Ruan, M.; de Heer, W.A.; First, P.N.; Stroscio, J.A. Observing the quantization of zero mass carriers in graphene. *Science* **2009**, *324*, 924–927. [CrossRef]

51. Ji, S.-H.; Hannon, J.; Tromp, R.; Perebeinos, V.; Tersoff, J.; Ross, F. Atomic-scale transport in epitaxial graphene. *Nat. Mater.* **2012**, *11*, 114–119. [CrossRef]

52. Yakes, M.K.; Gunlycke, D.; Tedesco, J.L.; Campbell, P.M.; Myers-Ward, R.L.; Eddy, C.R., Jr.; Gaskill, D.K.; Sheehan, P.E.; Laracuente, A.R. Conductance anisotropy in epitaxial graphene sheets generated by substrate interactions. *Nano Lett.* **2010**, *10*, 1559–1562. [CrossRef]

53. Janssen, T.J.B.M.; Tzalenchuk, A.; Lara-Avila, S.; Kubatkin, S.; Fal'ko, V.I. Quantum resistance metrology using graphene. *Rep. Prog. Phys.* **2013**, *76*, 104501. [CrossRef]

54. Kruskopf, M.; Pakdehi, D.M.; Pierz, K.; Wundrack, S.; Stosch, R.; Dziomba, T.; Götz, M.; Baringhaus, J.; Aprojanz, J.; Tegenkamp, C.; et al. Comeback of epitaxial graphene for electronics: Large-area growth of bilayer-free graphene on SiC. *2D Mater.* **2016**, *3*, 041002. [CrossRef]

55. Jobst, J.; Waldmann, D.; Speck, F.; Hirner, R.; Maude, D.K.; Seyller, T.; Weber, H.B. Transport properties of high-quality epitaxial graphene on 6H-SiC (0001). *Solid State Commun.* **2011**, *151*, 1061–1064. [CrossRef]

56. Tedesco, J.L.; VanMil, B.L.; Myers-Ward, R.L.; McCrate, J.M.; Kitt, S.A.; Campbell, P.M.; Jernigan, G.G.; Culbertson, J.C.; Eddy, C.R., Jr.; Gaskill, D.K. Hall effect mobility of epitaxial graphene grown on silicon carbide. *Appl. Phys. Lett.* **2009**, *95*, 122102. [CrossRef]

57. van der Pauw, L.J. A method of measuring specific resistivity and Hall effect of discs of arbitrary shape. *Philips Res. Rep.* **1958**, *13*, 1–9.

58. Geim, A.K.; Novoselov, K.S. The Rise of Graphene. In *Nanoscience and Technology: A Collection of Reviews from Nature Journals*; World Scientific: Singapore, 2010; pp. 11–19.

59. Lin, Y.-M.; Farmer, D.B.; Jenkins, K.A.; Wu, Y.; Tedesco, J.L.; Myers-Ward, R.L.; Myers-Ward, C.R.; Gaskill, D.K.; Dimitrakopoulos, C.; Avouris, P. Enhanced performance in epitaxial graphene FETs with optimized channel morphology. *IEEE Electron. Device Lett.* **2011**, *32*, 1343–1345. [CrossRef]

60. Kedzierski, J.; Hsu, P.-L.; Healey, P.; Wyatt, P.W.; Keast, C.L.; Sprinkle, M.; Berger, C.; de Heer, W.A. Epitaxial graphene transistors on SiC substrates. *IEEE T Electron. Dev.* **2008**, *55*, 2078–2085. [CrossRef]

61. Ytterdal, T.; Cheng, Y.; Fjeld, T.O. *Device Modeling for Analog 801 and RF CMOS Circuit Design*; John Wiley & Sons: Hoboken, NJ, USA, 2003.

62. Röhrl, J.; Hundhausen, M.; Speck, F.; Seyller, T. Strain and charge in epitaxial graphene on silicon carbide studied by raman spectroscopy. *Mater. Sci. Forum* **2010**, *645–648*, 603–606. [CrossRef]

63. Thomsen, C.; Reich, S. Double resonant Raman scattering in graphite. *Phys. Rev. Lett.* **2000**, *85*, 5214–5217. [CrossRef]

64. Ferrari, A.C.; Meyer, J.; Scardaci, V.; Casiraghi, C.; Lazzeri, M.; Mauri, F.; Piscanec, S.; Jiang, D.; Novoselov, K.; Roth, S.; et al. Raman spectrum of graphene and graphene layers. *Phys. Rev. Lett.* **2006**, *97*, 187401. [CrossRef]

65. Das, A.; Pisana, S.; Chakraborty, B.; Piscanec, S.; Saha, S.K.; Waghmare, U.V.; Novoselov, K.S.; Krishnamurthy, H.R.; Geim, A.K.; Ferrari, A.C.; et al. Monitoring dopants by Raman scattering in an electrochemically top-gated graphene transistor. *Nat. Nanotechnol.* **2008**, *3*, 210–215. [CrossRef]

66. Ouerghi, A.; Kahouli, A.; Lucot, D.; Portail, M.; Travers, L.; Gierak, J.; Penuelas, J.; Jegou, P.; Shukla, A.; Chassagne, T.; et al. Epitaxial graphene on cubic SiC (111)/Si (111) substrate. *Appl. Phys. Lett.* **2010**, *96*, 191910. [CrossRef]

67. Mueller, N.S.; Heeg, S.; Alvarez, M.P.; Kusch, P.; Wasserroth, S.; Clark, N.; Schedin, F.; Parthenios, J.; Papagelis, K.; Galiotis, C.; et al. Evaluating arbitrary strain configurations and doping in graphene with Raman spectroscopy. *2D Mater.* **2017**, *5*, 015016. [CrossRef]

68. Verhagen, T.; Drogowska, K.; Kalbac, M.; Vejpravova, J. Temperature-induced strain and doping in monolayer and bilayer isotopically labeled graphene. *Phys. Rev. B* **2015**, *92*, 125437. [CrossRef]

69. Shtepliuk, I.; Ivanov, I.G.; Pliatsikas, N.; Iakimov, T.; Jamnig, A.; Sarakinos, K.; Yakimova, R. Probing the uniformity of silver-doped epitaxial graphene by micro-Raman mapping. *Phys. B Condens. Matter* **2020**, *580*, 411751. [CrossRef]

70. Shtepliuk, I.; Ivanov, I.G.; Iakimov, T.; Yakimova, R.; Kakanakova-Georgieva, A.; Fiorenza, P.; Giannazzo, F. Raman probing of hydrogen-intercalated graphene on Si-face 4H-SiC. *Mater. Sci. Semicond. Process.* **2019**, *96*, 145–152. [CrossRef]

71. Pisana, S.; Lazzeri, M.; Casiraghi, C.; Novoselov, K.S.; Geim, A.K.; Ferrari, A.C.; Mauri, F. Breakdown of the adiabatic Born–Oppenheimer approximation in graphene. *Nat. Mater.* **2007**, *6*, 198–201. [CrossRef] [PubMed]

72. Sun, D.; Wu, Z.-K.; Divin, C.; Li, X.; Berger, C.; de Heer, W.A.; First, P.N.; Norris, T.B. Ultrafast relaxation of excited Dirac fermions in epitaxial graphene using optical differential transmission spectroscopy. *Phys. Rev. Lett.* **2008**, *101*, 157402. [CrossRef] [PubMed]

73. Hofmann, T.; Boosalis, A.; Kühne, P.; Herzinger, C.; Woollam, J.A.; Gaskill, D.; Tedesco, J.; Schubert, M. Hole-channel conductivity in epitaxial graphene determined by terahertz optical-Hall effect and midinfrared ellipsometry. *Appl. Phys. Lett.* **2011**, *98*, 041906. [CrossRef]

74. Armakavicius, N.; Bouhafs, C.; Stanishev, V.; Kühne, P.; Yakimova, R.; Knight, S.; Hofmann, T.; Schubert, M.; Darakchieva, V. Cavity-enhanced optical Hall effect in epitaxial graphene detected at terahertz frequencies. *Appl. Surf. Sci.* **2017**, *421*, 357–360. [CrossRef]

75. Mammadov, S.; Ristein, J.; Koch, R.J.; Ostler, M.; Raidel, C.; Wanke, M.; Vasiliauskas, R.; Yakimova, R.; Seyller, T. Polarization doping of graphene on silicon carbide. *2D Mater.* **2014**, *1*, 035003. [CrossRef]

76. Coletti, C.; Emtsev, K.V.; Zakharov, A.A.; Ouisse, T.; Chaussende, D.; Starke, U. Large area quasi-free standing monolayer graphene on 3C-SiC (111). *Appl. Phys. Lett.* **2011**, *99*, 081904. [CrossRef]

77. Grubišić-Čabo, A.; Monash University, Melbourne, Victoria, Australia. Personal communication, 2020.

78. Mathieu, C.; Lalmi, B.; Menteş, T.; Pallecchi, E.; Locatelli, A.; Latil, S.; Belkhou, R.; Ouerghi, A. Effect of oxygen adsorption on the local properties of epitaxial graphene on SiC (0001). *Phys. Rev. B* **2012**, *86*, 035435. [CrossRef]

79. Coletti, C.; Riedl, C.; Lee, D.S.; Krauss, B.; Patthey, L.; von Klitzing, K.; Smet, J.H.; Starke, U. Charge neutrality and band-gap tuning of epitaxial graphene on SiC by molecular doping. *Phys. Rev. B* **2010**, *81*, 235401. [CrossRef]

80. Moon, J.; Curtis, D.; Bui, S.; Marshall, T.; Wheeler, D.; Valles, I.; Kim, S.; Wang, E.; Weng, X.; Fanton, M. Top-gated graphene field-effect transistors using graphene on Si (111) wafers. *IEEE Electron. Device Lett.* **2010**, *31*, 1193–1195. [CrossRef]

81. Ristein, J.; Mammadov, S.; Seyller, T. Origin of doping in quasi-free-standing graphene on silicon carbide. *Phys. Rev. Lett.* **2012**, *108*, 246104. [CrossRef] [PubMed]

82. Kopylov, S.; Tzalenchuk, A.; Kubatkin, S.; Fal'ko, V.I. Charge transfer between epitaxial graphene and silicon carbide. *Appl. Phys. Lett.* **2010**, *97*, 112109. [CrossRef]

83. Speck, F.; Ostler, M.; Röhrl, J.; Jobst, J.; Waldmann, D.; Hundhausen, M.; Ley, L.; Weber, H.B.; Seyller, T. Quasi-freestanding graphene on SiC (0001). *Mater. Sci. Forum* **2010**, *645–648*, 629–632. [CrossRef]

84. De Heer, W.A.; Berger, C.; Wu, X.; Sprinkle, M.; Hu, Y.; Ruan, M.; Stroscio, J.A.; First, P.N.; Haddon, R.; Piot, B.; et al. Epitaxial graphene electronic structure and transport. *J. Phys. D Appl. Phys.* **2010**, *43*, 374007. [CrossRef]

85. Van Bommel, A.; Crombeen, J.; Van Tooren, A. LEED and Auger electron observations of the SiC (0001) surface. *Surf. Sci.* **1975**, *48*, 463–472. [CrossRef]

86. Emtsev, K.; Speck, F.; Seyller, T.; Ley, L.; Riley, J.D. Interaction, growth, and ordering of epitaxial graphene on SiC {0001} surfaces: A comparative photoelectron spectroscopy study. *Phys. Rev. B* **2008**, *77*, 155303. [CrossRef]

87. Seubert, A.; Bernhardt, J.; Nerding, M.; Starke, U.; Heinz, K. In situ surface phases and silicon-adatom geometry of the (2 × 2) C structure on 6H-SiC (0001⁻). *Surf. Sci.* **2000**, *454*, 45–48. [CrossRef]

88. Malard, L.; Pimenta, M.; Dresselhaus, G.; Dresselhaus, M. Raman spectroscopy in graphene. *Phys. Rep.* **2009**, *473*, 51–87. [CrossRef]

89. De Heer, W.A.; Berger, C.; Wu, X.; First, P.N.; Conrad, E.H.; Li, X.; Li, T.; Sprinkle, M.; Hass, J.; Sadowski, M.L.; et al. Epitaxial graphene. *Solid State Commun.* **2007**, *143*, 92–100. [CrossRef]

90. First, P.N.; de Heer, W.A.; Seyller, T.; Berger, C.; Stroscio, J.A.; Moon, J.-S. Epitaxial graphenes on silicon carbide. *MRS Bull.* **2010**, *35*, 296–305. [CrossRef]

91. Orlita, M.; Faugeras, C.; Grill, R.; Wysmolek, A.; Strupinski, W.; Berger, C.; de Heer, W.A.; Martinez, G.; Potemski, M. Carrier scattering from dynamical magnetoconductivity in quasineutral epitaxial graphene. *Phys. Rev. Lett.* **2011**, *107*, 216603. [CrossRef] [PubMed]

92. Lebedev, A.A.; Agrinskaya, N.V.; Beresovets, V.A.; Kozub, V.I.; Lebedev, S.P.; Sitnikova, A.A. Low temperature transport properties of multigraphene structures on 6H-SiC obtained by thermal graphitization: Evidences of a presence of nearly perfect graphene layer. *arXiv Preprint*, 2012; arXiv:1212.4272.

93. Hwang, E.; Adam, S.; Sarma, S.D. Carrier transport in two-dimensional graphene layers. *Phys. Rev. Lett.* **2007**, *98*, 186806. [CrossRef]

94. Jernigan, G.G.; VanMil, B.L.; Tedesco, J.L.; Tischler, J.G.; Glaser, E.R.; Davidson, A., III; Campbell, P.M.; Gaskill, D.K. Comparison of epitaxial graphene on Si-face and C-face 4H SiC formed by ultrahigh vacuum and RF furnace production. *Nano Lett.* **2009**, *9*, 2605–2609. [CrossRef]

95. Hu, Y.; Ruan, M.; Guo, Z.; Dong, R.; Palmer, J.; Hankinson, J.; Berger, C.; de Heer, W.A. Structured epitaxial graphene: Growth and properties. *J. Phys. D Appl. Phys.* **2012**, *45*, 154010. [CrossRef]

96. Dimitrakopoulos, C.; Grill, A.; McArdle, T.J.; Liu, Z.; Wisnieff, R.; Antoniadis, D.A. Effect of SiC wafer miscut angle on the morphology and Hall mobility of epitaxially grown graphene. *Appl. Phys. Lett.* **2011**, *98*, 222105. [CrossRef]

97. Gaskill, D.K. Epitaxial Graphene. In *Handbook of Crystal Growth*; Elsevier: Amsterdam, The Netherlands, 2015; pp. 755–783.

98. Novoselov, K.S.; Geim, A.K.; Morozov, S.; Jiang, D.; Katsnelson, M.I.; Grigorieva, I.; Dubonos, S.; Firsov, A.A. Two-dimensional gas of massless Dirac fermions in graphene. *Nature* **2005**, *438*, 197–200. [CrossRef] [PubMed]

99. Elias, D.; Gorbachev, R.; Mayorov, A.; Morozov, S.; Zhukov, A.; Blake, P.; Ponomarenko, L.; Grigorieva, I.; Novoselov, K.; Guinea, F.; et al. Dirac cones reshaped by interaction effects in suspended graphene. *Nat. Phys.* **2011**, *7*, 701–704. [CrossRef]

100. Mayorov, A.S.; Elias, D.C.; Mukhin, I.S.; Morozov, S.V.; Ponomarenko, L.A.; Novoselov, K.S.; Geim, A.; Gorbachev, R.V. How close can one approach the Dirac point in graphene experimentally? *Nano Lett.* **2012**, *12*, 4629–4634. [CrossRef]

101. Bolotin, K.I.; Sikes, K.J.; Jiang, Z.; Klima, M.; Fudenberg, G.; Hone, J.; Kim, P.; Stormer, H. Ultrahigh electron mobility in suspended graphene. *Solid State Commun.* **2008**, *146*, 351–355. [CrossRef]

102. Aristov, V.Y.; Molodtsova, O.V.; Chaika, A.N. Graphene synthesized on Cubic-SiC(001) in Ultrahigh Vacuum: Atomic and Electronic Structure and Transport Properties. In *Growing Graphene on Semiconductors*; Motta, N., Iacopi, F., Coletti, C., Eds.; Pan Stanford: Singapore, 2017; pp. 27–75.

103. Ago, H.; Ito, Y.; Mizuta, N.; Yoshida, K.; Hu, B.; Orofeo, C.M.; Tsuji, M.; Ikeda, K.-i.; Mizuno, S. Epitaxial chemical vapor deposition growth of single-layer graphene over cobalt film crystallized on sapphire. *ACS Nano* **2010**, *4*, 7407–7414. [CrossRef]

104. Fukidome, H.; Miyamoto, Y.; Handa, H.; Saito, E.; Suemitsu, M. Epitaxial growth processes of graphene on silicon substrates. *Jpn. J. Appl. Phys.* **2010**, *49*, 01AH03. [CrossRef]

105. Ouerghi, A.; Ridene, M.; Balan, A.; Belkhou, R.; Barbier, A.; Gogneau, N.; Portail, M.; Michon, A.; Latil, S.; Jegou, P.; et al. Sharp interface in epitaxial graphene layers on 3 C-SiC (100)/Si (100) wafers. *Phys. Rev. B* **2011**, *83*, 205429. [CrossRef]

106. Pradeepkumar, A.; Zielinski, M.; Bosi, M.; Verzellesi, G.; Gaskill, D.K.; Iacopi, F. Electrical leakage phenomenon in heteroepitaxial cubic silicon carbide on silicon. *J. Appl. Phys.* **2018**, *123*, 215103. [CrossRef]

107. Pradeepkumar, A.; Mishra, N.; Kermany, A.R.; Boeckl, J.J.; Hellerstedt, J.; Fuhrer, M.S.; Iacopi, F. Catastrophic degradation of the interface of epitaxial silicon carbide on silicon at high temperatures. *Appl. Phys. Lett.* **2016**, *109*, 011604. [CrossRef]

108. Schedin, F.; Geim, A.K.; Morozov, S.V.; Hill, E.; Blake, P.; Katsnelson, M.; Novoselov, K.S. Detection of individual gas molecules adsorbed on graphene. *Nat. Mater.* **2007**, *6*, 652–655. [CrossRef]

109. Kotsakidis, J.C.; Grubišić-Čabo, A.; Yin, Y.; Tadich, A.; Myers-Ward, R.L.; Dejarld, M.; Pavunny, S.P.; Currie, M.; Daniels, K.M.; Liu, C.; et al. Freestanding n-Doped Graphene via Intercalation of Calcium and Magnesium Into the Buffer Layer-SiC (0001) Interface. *arXiv Preprint*, 2020; arXiv:2004.01383.

110. Leenaerts, O.; Partoens, B.; Peeters, F. Adsorption of H_2O, NH_3, CO, NO_2, and NO on graphene: A first-principles study. *Phys. Rev. B* **2008**, *77*, 125416. [CrossRef]

111. Kazakova, O.; Panchal, V.; Burnett, T.L. Epitaxial graphene and graphene–based devices studied by electrical scanning probe microscopy. *Crystals* **2013**, *3*, 191–233. [CrossRef]

112. Giusca, C.E.; Panchal, V.; Munz, M.; Wheeler, V.D.; Nyakiti, L.O.; Myers-Ward, R.L.; Gaskill, D.K.; Kazakova, O. Water affinity to epitaxial graphene: The impact of layer thickness. *Adv. Mater. Interfaces* **2015**, *2*, 1500252. [CrossRef]

113. Munz, M.; Giusca, C.E.; Myers-Ward, R.L.; Gaskill, D.K.; Kazakova, O. Thickness-dependent hydrophobicity of epitaxial graphene. *Acs Nano* **2015**, *9*, 8401–8411. [CrossRef] [PubMed]

114. Panchal, V.; Giusca, C.E.; Lartsev, A.; Martin, N.A.; Cassidy, N.; Myers-Ward, R.L.; Gaskill, D.K.; Kazakova, O. Atmospheric doping effects in epitaxial graphene: Correlation of local and global electrical studies. *2D Mater.* **2016**, *3*, 015006. [CrossRef]

115. Yang, Y.; Brenner, K.; Murali, R. The influence of atmosphere on electrical transport in graphene. *Carbon* **2012**, *50*, 1727–1733. [CrossRef]

116. Pearce, R.; Iakimov, T.; Andersson, M.; Hultman, L.; Spetz, A.L.; Yakimova, R. Epitaxially grown graphene based gas sensors for ultra sensitive NO2 detection. *Sens. Actuators B Chem.* **2011**, *155*, 451–455. [CrossRef]

117. Kong, L.; Enders, A.; Rahman, T.S.; Dowben, P.A. Molecular adsorption on graphene. *J. Phys. Condens. Matter.* **2014**, *26*, 443001. [CrossRef]

118. Nomani, M.W.; Shishir, R.; Qazi, M.; Diwan, D.; Shields, V.; Spencer, M.; Tompa, G.S.; Sbrockey, N.M.; Koley, G. Highly sensitive and selective detection of NO2 using epitaxial graphene on 6H-SiC. *Sens. Actuators B Chem.* **2010**, *150*, 301–307. [CrossRef]

119. Khomyakov, P.; Giovannetti, G.; Rusu, P.; Brocks, G.; van den Brink, J.; Kelly, P.J. First-principles study of the interaction and charge transfer between graphene and metals. *Phys. Rev. B* **2009**, *79*, 195425. [CrossRef]

120. Nath, A.; Currie, M.; Boyd, A.K.; Wheeler, V.D.; Koehler, A.D.; Tadjer, M.J.; Robinson, Z.R.; Sridhara, K.; Hernandez, S.C.; Wollmershauser, J.A.; et al. In search of quantum-limited contact resistance: Understanding the intrinsic and extrinsic effects on the graphene—Metal interface. *2D Mater.* **2016**, *3*, 025013. [CrossRef]

121. DeJarld, M.; Campbell, P.M.; Friedman, A.L.; Currie, M.; Myers-Ward, R.L.; Boyd, A.K.; Rosenberg, S.G.; Pavunny, S.P.; Daniels, K.M.; Gaskill, D. Surface potential and thin film quality of low work function metals on epitaxial graphene. *Sci. Rep.* **2018**, *8*, 1–11. [CrossRef]

122. Yang, S.; Zhou, P.; Chen, L.; Sun, Q.; Wang, P.; Ding, S.; Jiang, A.; Zhang, D.W. Direct observation of the work function evolution of graphene-two-dimensional metal contacts. *J. Mater. Chem.* **2014**, *2*, 8042–8046. [CrossRef]

123. Tanabe, S.; Sekine, Y.; Kageshima, H.; Nagase, M.; Hibino, H. Carrier transport mechanism in graphene on SiC (0001). *Phys. Rev. B* **2011**, *84*, 115458. [CrossRef]

124. Daniels, K.M.; Jadidi, M.M.; Sushkov, A.B.; Nath, A.; Boyd, A.K.; Sridhara, K.; Drew, H.D.; Murphy, T.E.; Myers-Ward, R.L.; Gaskill, D.K. Narrow plasmon resonances enabled by quasi-freestanding bilayer epitaxial graphene. *2D Mater.* **2017**, *4*, 025034. [CrossRef]

125. Ohta, T.; El Gabaly, F.; Bostwick, A.; McChesney, J.L.; Emtsev, K.V.; Schmid, A.K.; Seyller, T.; Horn, K.; Rotenberg, E. Morphology of graphene thin film growth on SiC (0001). *New J. Phys.* **2008**, *10*, 023034. [CrossRef]

126. Hass, J.; Feng, R.; Li, T.; Li, X.; Zong, Z.; de Heer, W.; First, P.; Conrad, E.; Jeffrey, C.; Berger, C. Highly ordered graphene for two dimensional electronics. *Appl. Phys. Lett.* **2006**, *89*, 143106. [CrossRef]

127. Chen, J.-H.; Jang, C.; Adam, S.; Fuhrer, M.; Williams, E.D.; Ishigami, M. Charged-impurity scattering in graphene. *Nat. Phys.* **2008**, *4*, 377–381. [CrossRef]

128. Sarma, S.D.; Adam, S.; Hwang, E.; Rossi, E. Electronic transport in two-dimensional graphene. *Rev. Mod. Phys.* **2011**, *83*, 407. [CrossRef]

129. Speck, F.; Jobst, J.; Fromm, F.; Ostler, M.; Waldmann, D.; Hundhausen, M.; Weber, H.B.; Seyller, T. The quasi-free-standing nature of graphene on H-saturated SiC (0001). *Appl. Phys. Lett.* **2011**, *99*, 122106. [CrossRef]

130. Chen, J.-H.; Jang, C.; Xiao, S.; Ishigami, M.; Fuhrer, M.S. Intrinsic and extrinsic performance limits of graphene devices on SiO_2. *Nat. Nanotechnol.* **2008**, *3*, 206. [CrossRef] [PubMed]

131. Morozov, S.; Novoselov, K.; Katsnelson, M.; Schedin, F.; Elias, D.; Jaszczak, J.A.; Geim, A. Giant intrinsic carrier mobilities in graphene and its bilayer. *Phys. Rev. Lett.* **2008**, *100*, 016602. [CrossRef]

132. Bekyarova, E.; Itkis, M.E.; Ramesh, P.; Berger, C.; Sprinkle, M.; de Heer, W.A.; Haddon, R.C. Chemical modification of epitaxial graphene: Spontaneous grafting of aryl groups. *J. Am. Chem. Soc.* **2009**, *131*, 1336–1337. [CrossRef]

133. Giesbers, A.; Procházka, P.; Flipse, C. Surface phonon scattering in epitaxial graphene on 6 H-SiC. *Phys. Rev. B* **2013**, *87*, 195405. [CrossRef]

134. Riedl, C.; Coletti, C.; Iwasaki, T.; Zakharov, A.; Starke, U. Quasi-free-standing epitaxial graphene on SiC obtained by hydrogen intercalation. *Phys. Rev. Lett.* **2009**, *103*, 246804. [CrossRef]

135. Wong, S.L.; Huang, H.; Wang, Y.; Cao, L.; Qi, D.; Santoso, I.; Chen, W.; Wee, A.T.S. Quasi-free-standing epitaxial graphene on SiC (0001) by fluorine intercalation from a molecular source. *ACS Nano* **2011**, *5*, 7662–7668. [CrossRef]

136. Gierz, I.; Suzuki, T.; Weitz, R.T.; Lee, D.S.; Krauss, B.; Riedl, C.; Starke, U.; Höchst, H.; Smet, J.H.; Ast, C.R. Electronic decoupling of an epitaxial graphene monolayer by gold intercalation. *Phys. Rev. B* **2010**, *81*, 235408. [CrossRef]

137. Virojanadara, C.; Zakharov, A.; Yakimova, R.; Johansson, L.I. Buffer layer free large area bi-layer graphene on SiC (0 0 0 1). *Surf. Sci.* **2010**, *604*, L4–L7. [CrossRef]

138. Oliveira, M.H., Jr.; Schumann, T.; Fromm, F.; Koch, R.; Ostler, M.; Ramsteiner, M.; Seyller, T.; Lopes, J.M.J.; Riechert, H. Formation of high-quality quasi-free-standing bilayer graphene on SiC (0 0 0 1) by oxygen intercalation upon annealing in air. *Carbon* **2013**, *52*, 83–89. [CrossRef]

139. Joucken, F.; Henrard, L.; Lagoute, J. Electronic properties of chemically doped graphene. *Phys. Rev. Mater.* **2019**, *3*, 110301. [CrossRef]

140. Lartsev, A.; Yager, T.; Bergsten, T.; Tzalenchuk, A.; Janssen, T.M.; Yakimova, R.; Lara-Avila, S.; Kubatkin, S. Tuning carrier density across Dirac point in epitaxial graphene on SiC by corona discharge. *Appl. Phys. Lett.* **2014**, *105*, 063106. [CrossRef]

141. Peles-Lemli, B.; Kánnár, D.; Nie, J.C.; Li, H.; Kunsági-Máté, S. Some unexpected behavior of the adsorption of alkali metal ions onto the graphene surface under the effect of external electric field. *J. Phys. Chem. C* **2013**, *117*, 21509–21515. [CrossRef]

142. Ludbrook, B.; Levy, G.; Nigge, P.; Zonno, M.; Schneider, M.; Dvorak, D.; Veenstra, C.; Zhdanovich, S.; Wong, D.; Dosanjh, P.; et al. Evidence for superconductivity in Li-decorated monolayer graphene. *Proc. Natl. Acad. Sci. USA* **2015**, *112*, 11795–11799. [CrossRef]

143. Feng, T.; Xie, D.; Lin, Y.; Tian, H.; Zhao, H.; Ren, T.; Zhu, H. Unipolar to ambipolar conversion in graphene field-effect transistors. *Appl. Phys. Lett.* **2012**, *101*, 253505. [CrossRef]

144. Moon, J.; Curtis, D.; Hu, M.; Wong, D.; McGuire, C.; Campbell, P.; Jernigan, G.; Tedesco, J.; VanMil, B.; Myers-Ward, R.; et al. Epitaxial-graphene RF field-effect transistors on Si-face 6H-SiC substrates. *IEEE Electron. Device Lett.* **2009**, *30*, 650–652. [CrossRef]

145. Lin, Y.-M.; Dimitrakopoulos, C.; Jenkins, K.A.; Farmer, D.B.; Chiu, H.-Y.; Grill, A.; Avouris, P. 100-GHz transistors from wafer-scale epitaxial graphene. *Science* **2010**, *327*, 662. [CrossRef] [PubMed]

146. He, Z.; Yu, C.; Liu, Q.; Song, X.; Gao, X.; Guo, J.; Zhou, C.; Cai, S.; Feng, Z. High temperature RF performances of epitaxial bilayer graphene field-effect transistors on SiC substrate. *Carbon* **2020**, *164*, 435–441. [CrossRef]

147. Yu, C.; He, Z.; Li, J.; Song, X.; Liu, Q.; Cai, S.; Feng, Z. Quasi-free-standing bilayer epitaxial graphene field-effect transistors on 4H-SiC (0001) substrates. *Appl. Phys. Lett.* **2016**, *108*, 013102. [CrossRef]

148. Hwang, W.S.; Zhao, P.; Tahy, K.; Nyakiti, L.O.; Wheeler, V.D.; Myers-Ward, R.L.; Eddy, C.R., Jr.; Gaskill, D.K.; Robinson, J.A.; Haensch, W.; et al. Graphene nanoribbon field-effect transistors on wafer-scale epitaxial graphene on SiC substrates. *APL Mater.* **2015**, *3*, 011101. [CrossRef]

149. Bianco, F.; Perenzoni, D.; Convertino, D.; De Bonis, S.; Spirito, D.; Perenzoni, M.; Coletti, C.; Vitiello, M.; Tredicucci, A. Terahertz detection by epitaxial-graphene field-effect-transistors on silicon carbide. *Appl. Phys. Lett.* **2015**, *107*, 131104. [CrossRef]

150. Illarionov, Y.Y.; Banshchikov, A.G.; Polyushkin, D.K.; Wachter, S.; Knobloch, T.; Thesberg, M.; Mennel, L.; Paur, M.; Stöger-Pollach, M.; Steiger-Thirsfeld, A.; et al. Ultrathin calcium fluoride insulators for two-dimensional field-effect transistors. *Nat. Electron.* **2019**, *2*, 230–235. [CrossRef]

151. Giannazzo, F.; Schilirò, E.; Lo Nigro, R.; Roccaforte, F.; Yakimova, R. Atomic Layer Deposition of High-k Insulators on Epitaxial Graphene: A Review. *Appl. Sci.* **2020**, *10*, 2440. [CrossRef]

152. Sangwan, V.K.; Hersam, M.C. Electronic transport in two-dimensional materials. *Annu. Rev. Phys. Chem.* **2018**, *69*, 299–325. [CrossRef]

153. Liu, Z.; Ma, L.; Shi, G.; Zhou, W.; Gong, Y.; Lei, S.; Yang, X.; Zhang, J.; Yu, J.; Hackenberg, K.P.; et al. In-plane heterostructures of graphene and hexagonal boron nitride with controlled domain sizes. *Nat. Nanotechnol.* **2013**, *8*, 119. [CrossRef]

154. Takahashi, N.; Watanabe, K.; Taniguchi, T.; Nagashio, K. Atomic layer deposition of Y2O3 on h-BN for a gate stack in graphene FETs. *Nanotechnology* **2015**, *26*, 175708. [CrossRef]

155. Geng, D.; Dong, J.; Ang, L.K.; Ding, F.; Yang, H.Y. In situ epitaxial engineering of graphene and h-BN lateral heterostructure with a tunable morphology comprising h-BN domains. *NPG Asia Mater.* **2019**, *11*, 1–8. [CrossRef]

Review

Twistronics in Graphene, from Transfer Assembly to Epitaxy

Di Wu, Yi Pan * and Tai Min

Center for Spintronics and Quantum Systems, State Key Laboratory for Mechanical Behavior of Materials, Xi'an Jiaotong University, Xi'an 710049, China; wdxzcuyms@stu.xjtu.edu.cn (D.W.); min.tai@xjtu.edu.cn (T.M.)
* Correspondence: yi.pan@xjtu.edu.cn

Received: 31 May 2020; Accepted: 6 July 2020; Published: 8 July 2020

Abstract: The twistronics, which is arising from the moiré superlattice of the small angle between twisted bilayers of 2D materials like graphene, has attracted much attention in the field of 2D materials and condensed matter physics. The novel physical properties in such systems, like unconventional superconductivity, come from the dispersionless flat band that appears when the twist reaches some magic angles. By tuning the filling of the fourfold degeneracy flat bands, the desired effects are induced due to the strong correlation of the degenerated Bloch electrons. In this article, we review the twistronics in twisted bi- and multi-layer graphene (TBG and TMG), which is formed both by transfer assembly of exfoliated monolayer graphene and epitaxial growth of multilayer graphene on SiC substrates. Starting from a brief history, we then introduce the theory of flat band in TBG. In the following, we focus on the major achievements in this field: (a) van Hove singularities and charge order; (b) superconductivity and Mott insulator in TBG and (c) transport properties in TBG. In the end, we give the perspective of the rising materials system of twistronics, epitaxial multilayer graphene on the SiC.

Keywords: twistronics; twisted bilayer graphene; flat band; SiC

---

1. Introduction

Since the year 2004, graphene [1] has been a focus of condensed matter physics and materials science due to their superior physical properties, especially those showing potential applications in next generation electronics and optoelectronics. Recently, the twistronics, which refers to the studies of the electronic properties of two-dimensional layered structures with particular twist angles between the layers, has become a fast rising new branch of the field in regard to two dimensional (2D) material.

The fascinating research on 2D materials experienced three sequential stages in its development, each represented by a new degree of freedom in the sense of material structure. Stage 1 is represented by a real 2D system, e.g., freestanding graphene, which allows the discovery of various properties superior to the 3D or quasi-2D systems; stage 2 is represented by tailor-made 2D heterojunctions, which give rise to a variety of van der Waals stacking systems and stage 3 is represented by a twist angle between the neighboring layers, which facilitates the fine tuning of electronics structures and magically induces strong correlation phenomenon at "magic" twist angles, e.g., twistronics.

The concept "twistronics" was first introduced by S. Carr et al. in a theory paper [2] to discuss general approaches to calculate the electronic structure of arbitrarily twisted bilayer 2D materials. Prior to that, researchers had already worked on this topic and discovered or predicted a number of interesting properties of twisted bilayer graphene (TBG), for example, van Hove singularity [3–6], a possible flat band at "magic" angles [7–11]. The rise of twistronics was boosted in 2018, by Jarillo-Herrero's group's experimental discovery of superconductivity [12] and Mott insulator [13] on the graphene device with almost the exact "magic" twist angle, which was fabricated by the so called "tear and stack" transfer assembly technique, as demonstrated in Figure 1a. Many other interesting properties, like a topological Chern insulator [14–16], local charge order [17–20], as well as those of twisted multilayer 2D heterostructures [11,21–23], have been reported soon after. Currently this field has become one of the hottest in the frontier. However, if the twistronics would one day come to practical electronics devices, it is worth putting effort into realizing such a material system by more feasible methods, like epitaxial growth.

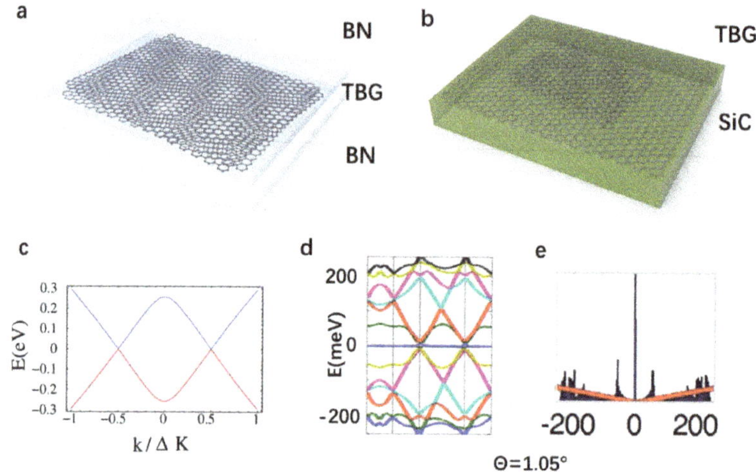

Figure 1. Schematic structure of twisted bilayer graphene (TBG) formed by transfer assembly (**a**) and epitaxial growth on SiC (**b**). The calculated band structure of TBG with θ = 3.9° (**c**) and θ = 1.05° (**d**), where flat bands (moiré) close to the Dirac point appears. (**e**) The density of states corresponding to θ = 1.05°. Panel (**c**) adapted with permission from Ref. [24], Copyright American Physical Society 2007. Panels (**d,e**) are adapted with permission from Ref. [8], Copyright PNAS 2011.

In this paper, we review the advances of this field, in the aspects of its band structure theory, transport properties and local electronic properties. Additionally, we will discuss the perspective of twistronics on epitaxial material systems, particularly epitaxial graphene on wide band semiconductor SiC (as in Figure 1b), aiming at promoting the application of twistronics in electronic devices based on the strong correlation effects, potentially useful in the next generation integrated circuits for quantum computing.

2. Electronic Structure of Twist Bilayer Graphene

When two layers of graphene are stacked up, they are in Bernal (AB) stacking or AA stacking when the crystal orientations of both layers are aligned, which is usually the case for grown crystals due to thermodynamics. However, when the bilayer structure is fabricated by transferring the exfoliated monolayers, the crystal orientations are mostly misaligned with a twist angle θ, which could be well controlled to the accuracy of 0.1° recently [25,26]. Such a twisted bilayer forms a super structure called the moiré pattern, with lattice constant $d = a\left(2\sin\left(\frac{\theta}{2}\right)\right)^{-1}$, where a = 0.246 nm, i.e., the lattice constant of graphene. In each moiré unit cell, there are three high symmetry spots, AA, AB and BA, named after the local stacking geometry.

A. H. Castro Neto and coworkers [24] studied the electronic structure of the such unique geometry of the bilayer graphene as early as in 2007. Within the framework of the continuum approximation on a small twist angle model, they found that the linear dispersion at low energy and the Dirac cones structure remain in the twisted bilayer, as shown in Figure 1c, but with a significant reduction of the

Fermi velocity for very small twist angles. Such results imply the twist angle could be a significant parameter that has a profound effect on the electronic structure of the bilayer graphene.

A theoretical breakthrough was made by R. Bistritzer and A. H. MacDonald in 2011 [8]. They unambiguously pointed out that the Fermi velocity at the Dirac point vanishes at a series of magic twist angles, e.g., $\theta = 1.05°$, when the dispersionless flat bands (also called moiré bands) are formed around the charge neutrality point, as shown in Figure 1d. Such a flat band contributes to a sharp peak to the density of state, which is shown in Figure 1e. They also remarked that the electron–electron interaction in the magic twist angle systems could give rise to strong correlation effects. E. Suárez Morell et al. [7] mentioned the link between superconductivity and the flat bands of twisted graphene, although they give the critical angle of 1.5° that is a little deviated from the magic angle.

From the view of the band structure, comparing with gapless monolayer graphene, it is two superlattice bandgaps that are next to the upper and bottom sub flat band due to the effect of interlayer hybridization in TBG. According to theoretical calculation [7,9,27], when the twist angle is increasing, and away from the first magic angle, the position of the gap would shift to higher energy until being disappeared as the two sub minibands become wider. Such twist angle tuned gaps are valuable for the application in the optoelectronic device, for example, B. Deng, et al. [28] demonstrated the gate-tunable photoresponse in the mid-infrared wavelength range of 5–12 μm in 1.81° TBG. Additionally, a number of recent optoelectronic works based on bigger twist angle TBG were done [29–32].

Another interesting feature in the electronic structure of the TBG is the doubled van Hove singularity in the density of states that generally exists at arbitrary angles TBG as a consequence of the saddle points at two emerging Dirac cones [6,9]. Additionally, a number of theory works have investigated the band topology [27,33–38] in the magic angle TBG system recently.

Although many of the interesting predicted properties are yet to be verified or investigated in detail by experiments, the flat bands have been directly observed in experiments, e.g., angle-resolved photoemission spectroscopy (ARPES) [39]. And the flat band related unconventional superconductivity and Mott insulating states have been discovered by Pablo Jarillo-Herrero's group [12,13] in 2018. In the following sections the major experimental achievements on TBG will be discussed.

3. Van Hove Singularities

Van Hove singularities (VHS) in the density of states are attractive properties of TBG, since strong correlation effects like superconductivity, ferromagnetism and the charge density wave could be induced when the system Fermi level is tuned to van Hove singularities energy levels. The VHS of TBG at low energy come from the saddle points formed when the Dirac cones of each graphene layer cross and merge together, as shown in Figure 2a [40]. Obviously, the VHS are always in a pair and are symmetric to the charge neutral point (CNP) in energy levels.

Actually, van Hove singularities are the first unique and well-studied phenomenon in TBG. The experimental studies on the moiré flat band, superconductivity, etc., are later inspired by the discovery of van Hove singularities. The earliest experimental work of the van Hove singularity in TBG are carried out by E. Y. Andrei's group, with scanning tunneling microscopy (STM) [3]. They have found a pair of sharp peaks symmetrical to the charge neutrality point in the scanning tunneling spectroscopy (STS) taken on the highly oriented pyrolytic graphite (HOPG) surface with the moiré pattern, as shown in Figure 2b,c. With the help of the theory, they could assign these peaks to the saddle points of the merged two Dirac cones, e.g., VHS levels. Later, they investigated the influence of the twist angle to the van Hove singularity energy levels by investigating chemical vapor deposition (CVD) grown TBG samples with various twist angles [4]. By measuring the Landau level positions under a high magnetic field, the Fermi velocity v_F could be deduced. It is found that the v_F of large twist angle TBG is similar to monolayer graphene, while the v_F of the small angle TBG is significantly lower, for example, $v_F = 0.87 \times 10^6$ m/s when $\theta = 1.16°$; $v_F = 1.1 \times 10^6$ m/s for monolayer graphene [41], as shown in Figure 2d.

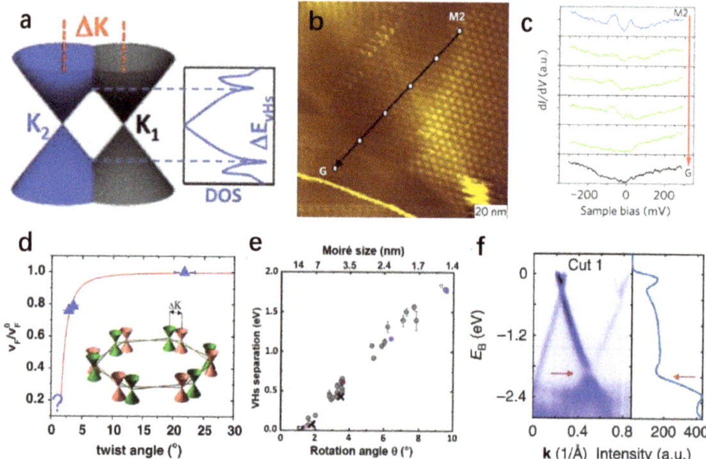

Figure 2. Van Hove singularities on the electronic structure of TBG. (**a**) Emergence of van Hove singularities (VHS) as a consequence of the rotation in reciprocal space. (**b**) Scanning tunneling microscopy (STM) image of the boundary of the moiré pattern. (**c**) Corresponding spatial dependence of tunneling spectra along a line connecting point M2 to G in (**b**). (**d**) Angle dependence of the Fermi velocity renormalization. Inset, Dirac cones of twisted layers. (**e**) VHS separation as a function of rotation angle θ. (**f**) μ-ARPES spectra of TBG. Red arrows indicate the minigap band topology arising from interlayer coupling. Panels (**a,e**) adapted with permission from Ref. [40], Copyright American Physical Society 2012; Panels (**b,c**) adapted with permission from Ref. [3], Copyright Nature Publishing Group 2010; Panel **d** adapted with permission from Ref. [4], Copyright American Physical Society 2011 and Panel (**f**) adapted with permission from Ref. [30], Copyright Nature Publishing Group 2016.

On the twisted multilayer graphene epitaxial grown on 6H-SiC(000$\bar{1}$), J.-Y. Veuillen's group systematically investigated the VHS levels by measuring the STS on samples with varying twist angles [40], as shown in Figure 2e. They have found that $\Delta E_{VHS} = 2hv_F \Gamma_K sin(\theta/2) - 2t_\theta$, where v_F is the Fermi velocity, Γ_K is the wave vector of the Dirac point in monolayer graphene and t_θ is the modulus of the amplitude of the main Fourier components of the interlayer potential for monolayer graphene. L. He's group [5] reported similar results on TBG samples grown on Rh foil, and reveal the coupling between epitaxial graphene and substrate could also influence the energy levels of VHS. These results are verified again by J. Yin et al. [30] on epitaxial 19.1° TBG on Cu foil by direct observation of the saddle point on the valance band of TBG by μ-ARPES, as shown in Figure 2f.

4. Mott Insulator and Superconductivity

The superconductivity in TBG with the magic angle was firstly reported by Y. Cao et al. in 2018 [12], and confirmed by many other researchers soon after [14,42,43]. The superconducting behavior of TBG is largely similar to the doped Mott insulator, similar to unconventional high T_C superconductor. Although it is well known that the monolayer graphene is a zero gap semiconductor with the highest mobility [44–46] among all the materials, Y. Cao et al. surprisingly discovered the insulating behavior in the TBG in 2016 [26]. They had successfully fabricated dual-gate hall bar devices (Figure 3a) with two TBG of different twist angles. The one with a twist angle > 3° shows a normal V-shape dip in the conductivity-carrier density curve, as shown in Figure 3b, which indicates the normal behavior of conductivity increases with carrier density around the zero density (charge neutral) point. However, in the small angle sample $\theta = (2.0 \pm 0.5)$ °, they observe two insulating states, which are symmetric on both sides of the charge neutrality point. Such insulating states are suppressed when the temperature increases. Interestingly, the total density of the states required to fill up the insulating gaps, $n = 7.5 \times 10^{12}$ cm^{-2}, is 4 times of the mini Brillouin zone area, indicating two fold valley degeneracy and two fold spin degeneracy. Therefore, they could derive the unit cell area of the moiré superlattice to be $4/n = 53.3$ nm^2, which correspond to the twist angle of $\theta = 1.8°$, matching the target value of $\theta = (2.0 \pm 0.5)°$.

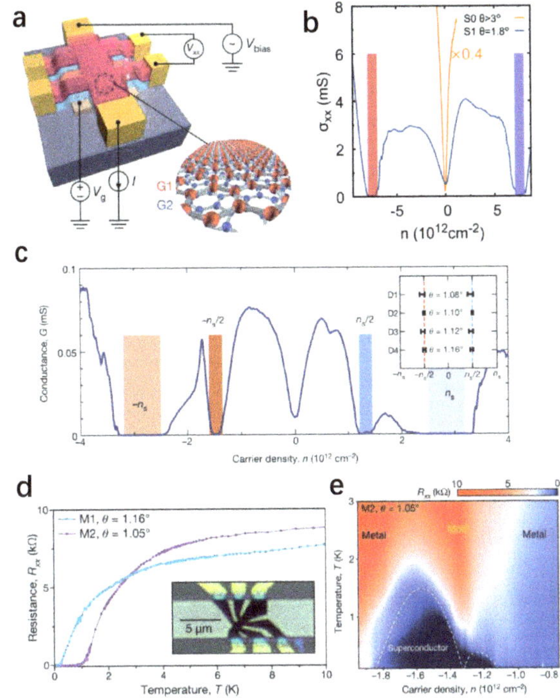

Figure 3. Mott insulator and superconductivity verified by the transport measurement in magic angle TBG. (**a**) Schematic of a typical TBG device and the four-probe measurement. (**b**) Comparison of the conductivity of two TBG devices. (**c**) Measured conductance of the TBG device D1 at T = 0.3 K. Inset, the carries density locations of half-filling states in the four different devices. (**d**) Longitudinal resistance measured in two devices respectively. Inset: an optical image of device M1. (**e**) Longitudinal resistance versus temperature and carries density for device M2. Panel (**a,d,e**) adapted with permission from Ref. [12], Copyright Nature Publishing Group 2018. Panel (**b**) adapted with permission from Ref. [26], Copyright American Physical Society 2016. Panel (**c**) adapted with permission from Ref. [13], Copyright Nature Publishing Group 2018.

The insulating behavior could be well explained by the flat band in magic angle TBG, as introduced in Section 1. Basically, when the Fermi level crosses the flat band, the dynamic energy of electron is limited by the width of the flat band. The Bloch electron in flat band would have a very large effective mass, and thus the Coulomb interaction between electrons would dominate their behavior, i.e., Mott insulator at low temperature. Additionally, due to the combination of valley degeneracy and spin degeneracy, the flat band is fourfold degeneracy.

In 2018, Y. Cao et al. [13] verified the Mott insulating states of TBG by measuring devices with twist angles carefully controlled to 1.1°, close to the first magic angle. The full filling condition is four electrons per moiré unit cell, corresponding to a carrier density n_s (in this work $n_s = 2.7 \times 10^{12}$ cm^{-2}). It was found that the insulating states not only appear at the charge neutral point (n = 0) and the full filling condition ($n = n_s$), but also at the half filling condition ($n = \pm n_s/2$), as shown in Figure 3c. Additionally, by measuring the Shubnikov–de Haas oscillation frequency f_{SdH}, they deduced that the degeneracy was to be 4 at charge neutrality and 2 at the half-filling states, and suggested the halved degeneracy of the Fermi pockets is related to the spin–charge separation. These results are unambiguous proof of Mott-like correlated behavior at half-filling. In a back to back paper on the same journal, Y. Cao et al. reported unconventional superconductivity in the same material system, as shown in Figure 3d, which was realized by back gate tuned electrostatic doping into the TBG slightly away from the correlated insulating state [12]. They observed tunable zero-resistance states, i.e., superconductivity, in two TBG devices with critical temperature of 1.7 K and 0.5 K respectively. Both devices show a maximum critical field of about 70 mT. The temperature and magnetic field–carrier density phase diagram demonstrate two dome-shaped superconducting phase regions approximate to both sides of the half-filling Mott-insulator region, as in Figure 3e. Such behavior of magic angle TBG is very similar to that of cuprates superconductors.

These works immediately inspired enormous attention to the TBG systems, both experimental and theoretical [14,42,47–59]. Due to the fourfold degeneracy of the moiré flat band, one can define a filling factor $v = n/n_0 = 4n/n_s$ (n_0 is the carrier density for one electron per moiré unit cell). It is expected that insulating states appears when v is the integer and superconducting exists between them. Indeed, X. Lu et al. reported that on a more accurate ($\theta = 1.1 \pm 0.02°$) magic angle TBG device, the insulating state appears at the integer filling factor positions, as shown in Figure 4a [14]. The temperature–carrier density phase diagram shows superconductivity up to 3 K around $v \approx -2$. Surprisingly, there are also three new superconducting domes at much lower temperatures, close to the $v = 0$ and $v = \pm 1$ insulating states. Moreover, at $v = -1$, when the perpendicular magnetic field is higher than 3.6 T, the insulating state exhibits a sharp hysteretic resistance and the longitudinal resistance saturated at about 13 kΩ (Figure 4b). These details imply orbital magnetism and a non-zero Chern number induced by a possible field-driven phase transition at low temperature. E. Codecido et al. [42] reported an insulating state at a filling factor as high as 12 in the TBG with a twist angle of 0.93°, as shown in Figure 4c,d, which is surprising since at such a high filling condition the Fermi level has reached the dispersing bands. M. Yankowitz et al. [56] induced superconductivity in TBG with twist angle $\theta = 1.27°$, larger than the first magic angle, by varying the interlayer spacing with hydrostatic pressure [60]. These superconducting (SC) domes are almost next to the correlated insulate states region in the B–T phase diagram, similar to the effect of the doping of the Mott insulator and implying the strong correlation effect between the electrons on the flat band in special filling. However, some theoretical works attributed the superconductivity in TBG to phonon driven [48,51].

In the above cases, the "tear and stack" technique were adopted by most of the groups to the fabricated magic angle twisted bilayer graphene device. This method [25,26] utilizes the van der Waals force between different materials to transfer graphene on different substrates. Precisely controlling the twist angle in the method is the key to realize the desired magic angle twisted bilayer graphene device. This is controlled by an accurate micro-manipulation stage and monitored by optical microscopy for confirming the twist angle in situ. Besides, in Y. Cao and X. Lu's case, they both rotated the separated graphene pieces purposefully by an angle slightly larger than the desired twist angle, owing to the high risk of relaxation of the twist angle to random lower values. Furthermore, X. Lu, et al. further carried out a mechanical cleaning process for squeezing the trapped blister out and releasing the local strain. More details and other different synthetic methods of twisted graphene can refer to this review paper [61], which introduced the method more comprehensively.

Moreover, in the early transport measurement of TBG devices, the filling is mainly tuned by the back gate, as shown in Figure 3a. One should note that the back gate not only changes the carrier density, but also changes the electric displacement field. Theory suggests the vertical electrical field would influence the band structure of the bilayer graphene. To eliminate this effect, researchers normally employ a symmetric due gate design for the devices. More details of the physics in such systems are still yet to be further investigated on more sophisticated devices.

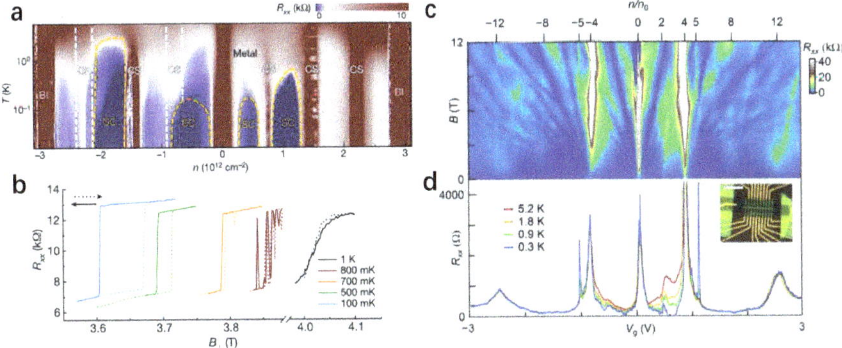

Figure 4. Tunable superconducting and insulate state in TBG. (**a**) Color plot of longitudinal resistance versus carrier density and temperature. SC: Superconducting state; CS: correlated state; BI: band insulator. (**b**) Longitudinal resistance plotted versus magnetic field at various temperatures. (**c**) Longitudinal resistance versus magnetic field and filling factor up to ±12. (**d**) Longitudinal resistance versus back gate voltage at various temperatures. Inset: an optical image of the TBG device with a scale bar of 10 μm. Panel (**a,b**) adapted with permission from Ref. [14], Copyright Nature Publishing Group 2019. Panel (**c,d**) adapted with permission from Ref. [42], Copyright AAAS 2019.

5. Ferromagnetism and Quantum Anomalous Hall Effect

The ferromagnetic state in TBG has been predicted by L. A. Gonzalez-Arraga et al. [62] in 2017 and antiferromagnetism was suggested as well [63]. Basically, fine tuning of the band filling would lift the degeneracy, and thus induce ferromagnetism in the system. For example, the spin symmetry and valley symmetry could be partially or completely broken in the 1/4, 1/2 and 3/4 filling. A. L. Sharpe et al. reported that the transport measurement shows hysteretic behavior with respect to an applied out-of-plane magnetic field B below 3.9 K, near the 3/4 filling of the moiré band in a TBG Hall bar device [64], as shown in Figure 5a. Considering the absence of transition metal and other heavy elements, the large amorous hall signal in TBG is surprising. This suggests spin and valley symmetries in the correlated insulated states are spontaneously broken. M. Serlin et al. also found the ferromagnetic state near the 3/4 filling in TBG devices [16]. The Hall resistivity is hysteretic, with a critical field of several tens of mT. At low temperature, lateral resistance R_{yx} quickly saturated at quantum resistivity h/e^2 along with a low longitudinal resistivity $R_{xx} < 1$ kΩ, as shown in Figure 5b, which reveals a topological insulating state with Chern number = 1, similarly to the result by X. Lu et al. [14]. These results again indicate a quantum anomalous Hall (QAH) state by spontaneously broken time-reversal symmetry of the moiré band. Additionally, they both found the quantum anomalous Hall state can be switched on/off by a low direct current [16,64], as shown in Figure 5c,d, providing a possibility for functional devices.

This unusual magnetism in TBG, revealed by the anomalous Hall effect, cannot be explained by a trivial reason like adsorbed impurities or defects. Due to the absence of the *d* electron or *f* electron, the unusual magnetism at the 1/4 filling can be attributed to a field stabilized orbital magnetic state [14,65]. Additionally, the certainly potential ferromagnetic topological insulator, found at the 3/4 filling, is explained by spin or valley symmetry broken from the flat band. A series of theoretical works have attempted to find the reason of the electrically controllable magnetism [58,66–69] and the origin of the Chern number in TBG by calculating the topology of the flat band [34,35,37]. To better understand the magnetism experimentally, these spin or valley polarized states, which are around the AA region as suggested by the theory, need to be verified by performing more detailed local probe characterization.

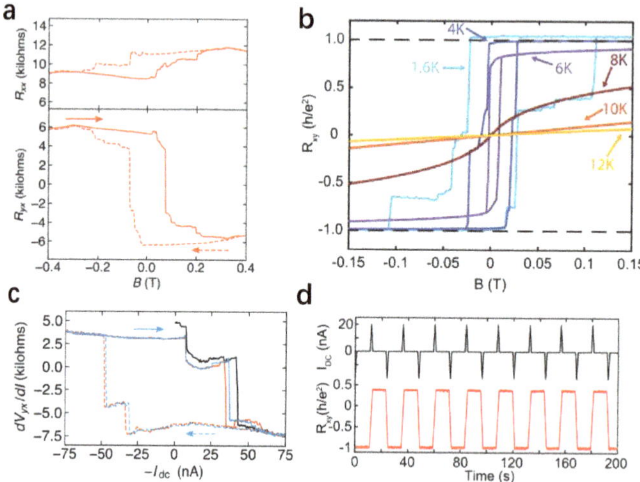

Figure 5. Ferromagnetism and quantum anomalous Hall effect (QAHE) in TBG. (**a**) Magnetic field dependence of the longitudinal resistance (upper panel) and Hall resistance (lower panel) with $n/n_s = 0.746$ at 30 mK. (**b**) Hall resistance as a function of magnetic field at various temperatures with $n/n_s \approx 3/4$. (**c**) Differential Hall resistance tuned by a DC bias with $n/n_s = 0.746$ at 2.1 K. (**d**) Nonvolatile electrical writing and reading of a magnetic bit at $T = 6.5$ K and $B = 0$. Panel (**a,c**) adapted with permission from Ref. [64], Copyright AAAS 2019. Panel (**c,d**) adapted with permission from Ref. [16], Copyright AAAS 2020.

6. Local Probe Measurement of TBG

Although the transport measurement already revealed many fascinating properties of TBG systems, one should note that the published results sometime are puzzling or contradicting to each other, for example, different determination to the insulating states at a certain filling to the band [13,42,64]; disagreed positions of the superconducting states in the phase diagram [12,14] and different critical magnetic field or critical temperature values [12,56]. These contradictions are mainly attributed to the unavoidable variation in device fabrication, local inhomogeneity or twist angle disorder. To avoid these deviations in the measurement that are carried out on a relatively large area, a local probe measurement that could zoom in to the individual moiré unit cell would be very helpful.

Due to the atomic resolution STM measurements on TBG, L. Yin et al. [70] observed the one-dimensional conducting channels locating at the edges of the domain wall between the AB and BA regions; S. Huang et al. [71] observed a double-line network of enhanced density of states on TBG with the moiré unit cell lattice constant of about 50 nm, as shown in Figure 6a. The density of the state is enhanced only when the gap was opened by applying a positive electric field, suggesting topologically protected helical edge states on the network. On the TBG samples with varying twist angles from 0.79° to 3.48°, A. Kerelsky et al. [19] detected the VHS features in STS spectra. They also found that the half-width of each individual VHS is minimized at the magic angle, as shown in Figure 6b, which means the electron correlations are maximized.

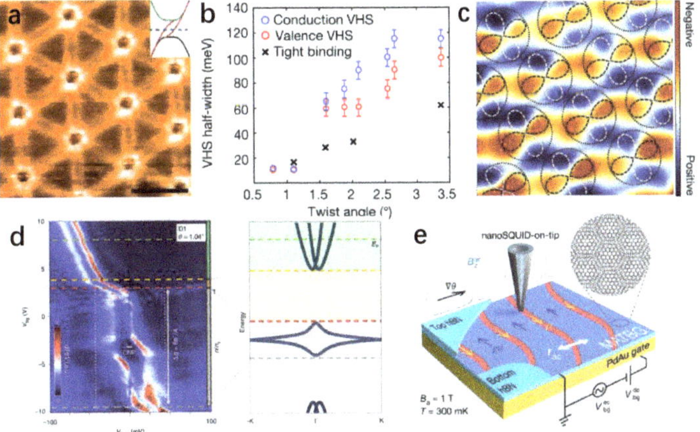

Figure 6. Local probe measurements of TBG. (**a**) STM density of states (DOS) mapping of the moiré pattern with period —50 nm. Inset: scheme of the band structure. (**b**) Experimental conduction and valence VHS half-widths and tight-binding half-widths versus twist angle. (**c**) Superimposed STM DOS map showing the stripe net charge order. (**d**) Evolution of the TBG point spectrum with back-gate voltage. Right panel: schematic of the TBG band structure. (**e**) Schematic of SQUID on tip scanning over magic angle TBG. Panel (**a**) adapted with permission from Ref. [71], Copyright American Physical Society 2018. Panel (**b**) adapted with permission from Ref. [19], Panel (**c**) adapted with permission from Ref. [18] and Panel (**d**) adapted with permission from Ref. [17], Copyright Nature Publishing Group 2019. Panels (**e**) adapted with permission from Ref. [72], Copyright Nature Publishing Group 2020.

The influences of the electron–electron correlation interaction to the local electronic structure and local charge ordering are also very suitable for scanning probe measurement. By taking STS on the AA region of the moiré unit cell in TBG with the magic angle, Y. Xie et al. [20] studied the flat band peak position under different filling conditions. It was found that in the fully filled or empty condition, the shift of the flat band peak is proportional to the density of states (DOS) at the Fermi level, as shown in Figure 6d, indicating that the electron correlation is weak; while in the partial filling condition, an abrupt distortion of the spectra that cannot be explained by the mean-field model has been observed, indicating that a strong correlation effect is indeed playing a crucial role. Such a conclusion is also supported by the STS measurement on a similar system by Y. Choi et al. [17]. In addition, Y. Choi et al. observed an enhanced flat band peak splitting, which was attributed to exchange interactions. Furthermore, the strong correlation effect could also induce a stripe charge order, as reported by Y. Jiang et al. [18], as shown in Figure 6c.

Besides the scanning tunneling microscope, the nanoscale on-tip scanning superconducting quantum interference device (SQUID-on-tip) [73] is a powerful tool to study the local magnetic properties at high spatial resolution. Recently, A. Uri et al. [72] investigated the influence of local θ variation on the Landau levels in the quantum Hall states in magic angle TBG, at an ultra-high precision of 0.002° with SQUID-on-tip, as shown in Figure 6e. It was found that the correlated states being fragile with the twist-angle disorder, and the in-plane electron fields generated by the gradients of θ would affect the phase diagram of correlated superconducting states.

7. Twisted Multilayer Graphene

Apart from the gating as a major tool to tune the correlated states in TBG, the number of layers is also a knob for tuning. Actually, correlation effects also exist in the multilayer graphene with or without the twist angle, for example, the twisted double-bilayer graphene (TDBG) [21,74–77] formed between twisted AB stacking bilayer graphene. It is well known that the AB stacking bilayer graphene has parabolic bands touching at low energies close to the K point, and the gap opening at the charge neutral point could be induced under electric displacement fields, as shown in Figure 7a. Theory suggests that strong electron–electron interaction may occur in such a TDBG system at the 1/2 and 1/4 filling of the flat band [78].

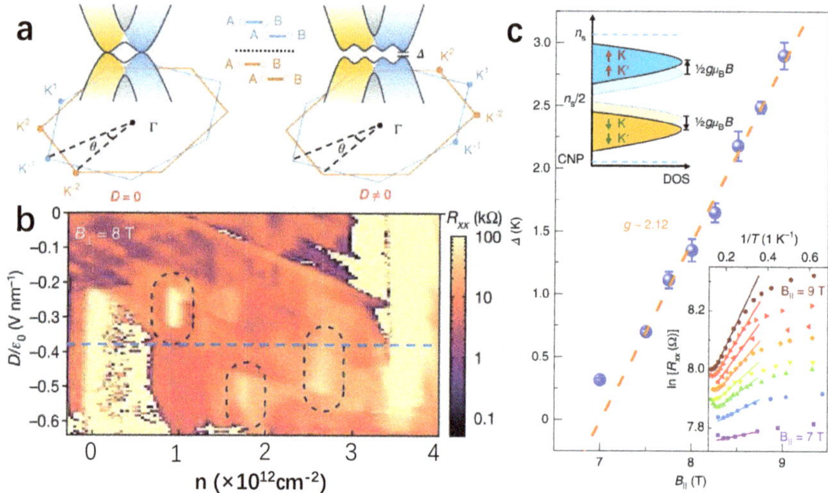

Figure 7. The correlated insulated and potential spin-polarized state in TBG. (**a**) Schematics of a moiré unit cell in TBG and the band structure with and without tuning by the displaced field. (**b**) Longitudinal resistance R_{xx} plot versus the displaced field and carrier density for the TDBG device in magnetic fields of $B_\perp = 8$ T. The dashed blue line denoted the original center of the insulated state with the 1/4 and 1/2 filling at $B = 0$. (**c**) Thermal activation gap Δ of the $n_s/2$ insulated state as a function of $B_\parallel$. Top inset: the single-particle flat band is split into upper and lower spin-polarized many-body bands by electron–electron interactions. Bottom inset: the fitting of R–T by $R \approx \exp(\Delta/2kT)$ for extracting the gap. Panel (**a,c**) adapted with permission from Ref. [74], Copyright Nature Publishing Group 2020. Panel (**b**) adapted with permission from Ref. [75], Copyright Nature Publishing Group 2020.

Indeed, a number of experimental works show evidence that the correlated insulate states exist in TDBG. Y. Cao et al. recently investigated the response of the correlated states to magnetic fields, perpendicular or in-plane [75]. At $B_\perp = 8$ T, the correlated insulate state at 1/2 n_s shifts to a bigger D region in the n-D map, and the 1/4 n_s insulate state shifts to a lower D region, as shown in Figure 7b. At the paralleled magnetic field, no such strong shift is observed at all fractional filling. In the evolution of the thermally activated gap of the 1/2 n_s state, they find a g-factor of $g_\parallel \approx 1.5$ for the in-plane magnetic field, closer to $g = 2$ that corresponds to electron spin contribution, rather than $g_\perp \approx 3.5$ for the out of plane magnetic field. These results suggest that the correlated states are spin-polarized. C. Shen et al. obtained an effective g factor of 2.12 by fitting the thermal activation gap according to the Zeeman effect in in-plane magnetic field [74], as shown in Figure 7c, which indicates spin polarization at half-filling. X. Liu et al. also reported the ferromagnetic order and superconductivity at 1/2 filling in TDBG [77]. Additionally, they surprisingly observed that the superconductivity is enhanced in the low in-plane magnetic field region, revealed by a 50% increase of critical temperature as $B_\parallel$ increases from 0 to about 0.3 T, again suggesting unusual spin-polarized electron pairing [77].

On an ABC stacking trilayer graphene with the 1/4 or 1/2 filling of the flat band, G. Chen et al. observed the Mott-insulating state [79], superconductivity [80], ferromagnetism and Chern insulator [15], as shown in Figure 8a–c. By applying the vertical field, they observed splitting of the low energy band and suppressing of the high energy flat band, indicating correlation due to

Coulomb repulsion. L. Yin et al. reported Landau level spectroscopy measurements of ABC-stacked twisted trilayer graphene on graphite, and observed an approximately linear magnetic-field scaling of valley splitting and spin splitting, as shown in Figure 8d, which is attributed to a strong many-body correlation effect [81]. Additionally, by increasing the number of layers and complexity of the stack order, more twisted multilayer graphene come up in experiment [22,82,83] and theory [23,65,84–86], thus more fascinating properties are yet to be explored.

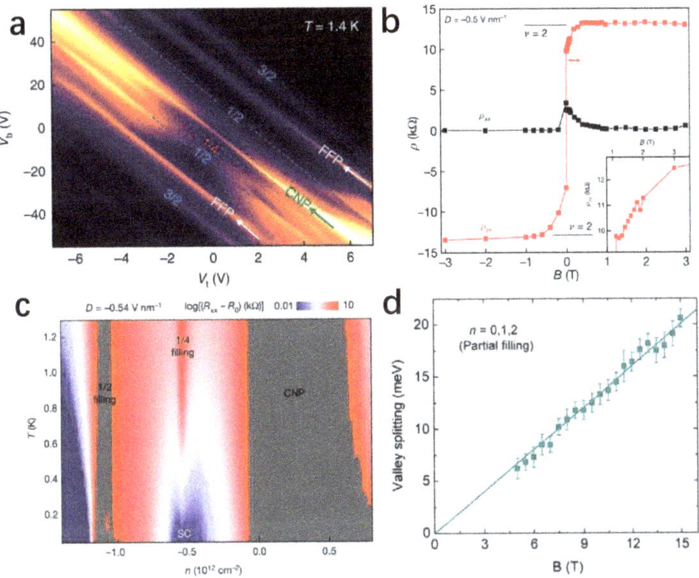

Figure 8. Correlated insulating, superconducting state and quantum anomalous hall effect (QAHE) in ABC-trilayer graphene. (**a**) Longitudinal resistance versus bottom gate voltage and tap gate voltage. (**b**) ρ_{xx} and ρ_{yx} versus the magnetic field with $D = -0.5$ V nm^{-1}. Inset: a zoomed-in plot of ρ_{yx} at a small magnetic field. (**c**) Color plot of the normalized resistance versus carrier density and temperature at $D = -0.54$ V nm^{-1}. (**d**) Valley splitting versus the magnetic field for partially filled at a low Landau level (LL) factor $n = 0, 1, 2$. Panel (**a**) adapted with permission from Ref. [79], Copyright Nature Publishing Group 2019. Panel (**b**) adapted with permission from Ref. [80], Copyright Nature Publishing Group 2019. Panel (**c**) adapted with permission from Ref. [15], Copyright Nature Publishing Group 2020. Panel (**d**) adapted with permission from Ref. [81], Copyright American Physical Society 2019.

8. Twistronics in Epitaxial Graphene on SiC

So far, the materials for twistronics studies are dominated by transfer assembly of exfoliated graphene, mono- and bi-layer. Due to its great precision, the tear and stack method [25,87,88] of transfer assembly enables the successful fabrication of desired magic angle TBG devices. On the other hand, future application of twistronics would demand mass production of TBG based devices, and thus require more efficient and economically available method for the production of TBG materials. The epitaxial graphene on wide band semiconductor SiC stands out as a very promising candidate, due to its natural compatibility with the semiconductor industry. Therefore, more attention should be paid to SiC supported epitaxial graphene. (See Figure 9)

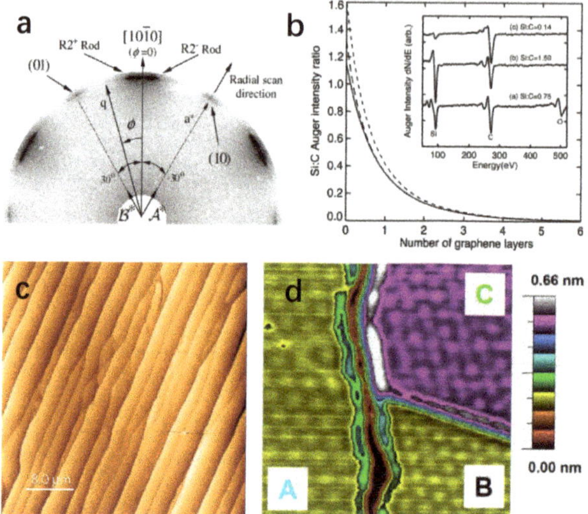

Figure 9. Twisted epitaxial graphene on SiC. (**a**) LEED image acquired at 67.9 eV from 4H-SiC(000$\bar{1}$) with 10 layers of graphene, showing only graphene spots and diffuse arcs. (**b**) Auger peak intensity ratio versus number of graphene layers on SiC(0001) substrates. Inset: Auger spectra obtained after different process. (**c**) Atomic force microscope (AFM) image of graphene on 6H–SiC(0001) with a nominal thickness of 1.2 ML formed by annealing in Ar. (**d**) STM image showing three domains with different thicknesses and twist angles. Zone A: bilayer, $\theta = 14°$; zone B: 2-3 layers, $\theta = 6.0°$; zone C: 3 layers, $\theta = 6.0°$. Panel (**a**) adapted with permission from Ref. [89], Copyright American Physical Society 2008. Panel (**b**) adapted with permission from Ref. [90], Copyright Elsevier 2007. Panel (**c**) adapted with permission from Ref. [91], Copyright AAAS 2006. Panel (**d**) adapted with permission from Ref. [92], American Physical Society 2015.

As early as in 1974, A. J. Van Bommel, et al. already reported the ring shape Low Energy Electron Diffraction (LEED) pattern on graphitized SiC(000$\bar{1}$) [93], which now we know is multi-layer graphene with rotational domains of varying orientations. M. Naitoh et al. also observed many large and small domains with various periodicities at 6H-SiC(000$\bar{1}$) surfaces [94], which can be explained as moiré patterns formed on two graphene layers with different twist angles. On the 4H-SiC(000$\bar{1}$) substrate, J. Hass, et al. reported that multilayer graphene grown on the carbon terminated SiC surface contains three rotational domains, rotated by the 30 ± 2.2° direction, characterized by LEED as shown in Figure 9a, and these rotational stacking faults could be largely tuned by the annealing temperature [89].

Since the pioneer work by the W. A. de Heer group in 2004 [95], lots of attention has been focused on the SiC(0001), i.e., Si terminated face, due to the high quality of the self-limited epitaxial graphene up to two layers. Meanwhile, many surface science studies, by means of auger electron spectroscopy (AES) [90,96], XPS [97–99], LEEM [100], STM [101], etc., were carried out on the epitaxial graphene on the SiC(000$\bar{1}$), i.e., C terminated face. It was confirmed that the C-face could provide multilayer epitaxial graphene with a thickness up to 10 s of nanometers, as shown in Figure 9b. The precise number of layers could be determined in situ [102]. Additionally, the technique to control the thickness and domain size of C-face graphene has also been developed. K.V. Emtsev et al. [91], J. L. Tedesco et al. [103] and G. R. Yazdi et al. [104] reported the Ar atmosphere annealing method that could provide micrometer sized single domains, as shown in Figure 9c. The epitaxial graphene on C-face SiC tends to exhibit a variety of domain with various twist angles, as shown in Figure 9d, which could provide rich research objects simultaneously.

In recent years, the physics regarding twisted epitaxial graphene on SiC started to attract interest. Since 2012, the VHS has been systematically studied on the twisted multilayer epitaxial graphene grown on the C-face SiC [40,92,105]. In 2018, D. Marchenko, et al. [106] observed a very intense and very flat band on bilayer graphene on graphene epitaxial on 6H-SiC. W. Wang et al. also verified the existence of a nearly dispersionless electronic band near the Fermi level on 3C-SiC by angle resolved photoelectron spectroscopy [107].

Although the precision control of the twist angle of epitaxial graphene remains a technique problem yet to be solved, the epitaxial graphene on SiC already becomes a promising platform for the study as well as application of twistronics in graphene. We suggest more twistronics studies on this system in the aspects including but not limited to the transport measurement, local electronic structure and functional devices.

9. Conclusions and Outlooks

In conclusion, the twist angle, as a new degree of freedom to tune the electronic structure of bi- and multi-layer 2D materials, has opened up a door to the zoo of twistronics, i.e., various intriguing physical properties of magic twist angle systems, especially the magic angle bilayer graphene that give rise to strong electron–electron correlation effects. The origin of them are the moiré flat bands formed above and below the charge neutrality point when the twist angles are in a series of "magic" numbers, 1.08°, 0.5°, 0.35°, etc. Indeed, the existence of flat bands and van Hove singularity has been confirmed by the momentum space and real spaces electronic structure measurements. Strong correlation effects, like Mott insulator states, unconventional superconductivity, ferromagnetism and charge/spin order, has been reported by many groups via sophisticated band filling tuning in transport or local probe measurements.

On the other hand, since such a new field is still in its early stage, more works are still highly desired for both fundamental research and applications. A virgin land is the twistronics on other magic angle twisted 2D material beyond graphene, especially the various 2D semiconductors, like transition metal dichalcogenides, phosphorene, etc. For the future application of twistronics in single devices and integrated circuits, graphene family materials are the most promising candidates so far, due to its quantum effects that might be valuable for quantum information technology.

The epitaxial graphene on SiC substrates would become a major platform for twistronics due to the advantage of both the twist angle tenability and compatibility to industrial semiconductor technology. Transfer assembly technology has enabled the tremendous physics discoveries in twistronics, while epitaxial growth would open the way for its application.

Author Contributions: Writing—original draft preparation, D.W.; writing—review and editing, D.W. and Y.P.; project administration, Y.P.; funding acquisition, Y.P. and T.M. All authors have read and agreed to the published version of the manuscript.

Funding: This work was funded by the National Key R&D Program of China (2017YFA0206202), the Strategic Priority Research Program of the Chinese Academy of Sciences (Grant No. XDB30000000) and the National Science Foundation of China (11704303).

Acknowledgments: The authors acknowledge Hong Lei for useful discussions.

Conflicts of Interest: The authors declare no conflict of interest.

References

1. Novoselov, K.S.; Geim, A.K.; Morozov, S.V.; Jiang, D.; Zhang, Y.; Dubonos, S.V.; Grigorieva, I.V.; Firsov, A.A. Electric field effect in atomically thin carbon films. *Science* **2004**, *306*, 666–669. [CrossRef] [PubMed]
2. Carr, S.; Massatt, D.; Fang, S.; Cazeaux, P.; Luskin, M.; Kaxiras, E. Twistronics: Manipulating the electronic properties of two-dimensional layered structures through their twist angle. *Phys. Rev. B* **2017**, *95*, 075420. [CrossRef]
3. Li, G.; Luican, A.; Lopes dos Santos, J.M.B.; Castro Neto, A.H.; Reina, A.; Kong, J.; Andrei, E.Y. Observation of Van Hove singularities in twisted graphene layers. *Nat. Phys.* **2010**, *6*, 109–113. [CrossRef]
4. Luican, A.; Li, G.; Reina, A.; Kong, J.; Nair, R.R.; Novoselov, K.S.; Geim, A.K.; Andrei, E.Y. Single-Layer Behavior and Its Breakdown in Twisted Graphene Layers. *Phys. Rev. Lett.* **2011**, *106*, 126802. [CrossRef]

5. Yan, W.; Liu, M.; Dou, R.-F.; Meng, L.; Feng, L.; Chu, Z.-D.; Zhang, Y.; Liu, Z.; Nie, J.-C.; He, L. Angle-Dependent van Hove Singularities in a Slightly Twisted Graphene Bilayer. *Phys. Rev. Lett.* **2012**, *109*, 126801. [CrossRef] [PubMed]

6. De Laissardiere, G.T.; Mayou, D.; Magaud, L. Numerical studies of confined states in rotated bilayers of graphene. *Phys. Rev. B* **2012**, *86*, 125413. [CrossRef]

7. Suarez Morell, E.; Correa, J.D.; Vargas, P.; Pacheco, M.; Barticevic, Z. Flat bands in slightly twisted bilayer graphene: Tight-binding calculations. *Phys. Rev. B* **2010**, *82*, 121407. [CrossRef]

8. Bistritzer, R.; MacDonald, A.H. Moiré bands in twisted double-layer graphene. *Proc. Natl. Acad. Sci. USA* **2011**, *108*, 12233–12237. [CrossRef]

9. Lopes dos Santos, J.M.B.; Peres, N.M.R.; Castro Neto, A.H. Continuum model of the twisted graphene bilayer. *Phys. Rev. B* **2012**, *86*, 155449. [CrossRef]

10. Javvaji, S.; Sun, J.-H.; Jung, J. Topological flat bands without magic angles in massive twisted bilayer graphenes. *Phys. Rev. B* **2020**, *101*, 125411. [CrossRef]

11. Haddadi, F.; Wu, Q.; Kruchkov, A.J.; Yazyev, O.V. Moiré Flat Bands in Twisted Double Bilayer Graphene. *Nano Lett.* **2020**, *20*, 2410–2415. [CrossRef] [PubMed]

12. Cao, Y.; Fatemi, V.; Fang, S.; Watanabe, K.; Taniguchi, T.; Kaxiras, E.; Jarillo-Herrero, P. Unconventional superconductivity in magic-angle graphene superlattices. *Nature* **2018**, *556*, 43–50. [CrossRef] [PubMed]

13. Cao, Y.; Fatemi, V.; Demir, A.; Fang, S.; Tomarken, S.L.; Luo, J.Y.; Sanchez-Yamagishi, J.D.; Watanabe, K.; Taniguchi, T.; Kaxiras, E.; et al. Correlated insulator behaviour at half-filling in magic-angle graphene superlattices. *Nature* **2018**, *556*, 80–84. [CrossRef]

14. Lu, X.; Stepanov, P.; Yang, W.; Xie, M.; Aamir, M.A.; Das, I.; Urgell, C.; Watanabe, K.; Taniguchi, T.; Zhang, G.; et al. Superconductors, orbital magnets and correlated states in magic-angle bilayer graphene. *Nature* **2019**, *574*, 653–657. [CrossRef] [PubMed]

15. Chen, G.; Sharpe, A.L.; Fox, E.J.; Zhang, Y.H.; Wang, S.; Jiang, L.; Lyu, B.; Li, H.; Watanabe, K.; Taniguchi, T.; et al. Tunable correlated Chern insulator and ferromagnetism in a Moiré superlattice. *Nature* **2020**, *579*, 56–61. [CrossRef]

16. Serlin, M.; Tschirhart, C.L.; Polshyn, H.; Zhang, Y.; Zhu, J.; Watanabe, K.; Taniguchi, T.; Balents, L.; Young, A.F. Intrinsic quantized anomalous Hall effect in a Moiré heterostructure. *Science* **2020**, *367*, 900–903. [CrossRef]

17. Choi, Y.; Kemmer, J.; Peng, Y.; Thomson, A.; Arora, H.; Polski, R.; Zhang, Y.; Ren, H.; Alicea, J.; Refael, G.; et al. Electronic correlations in twisted bilayer graphene near the magic angle. *Nat. Phys.* **2019**, *15*, 1174–1180. [CrossRef]

18. Jiang, Y.; Lai, X.; Watanabe, K.; Taniguchi, T.; Haule, K.; Mao, J.; Andrei, E.Y. Charge order and broken rotational symmetry in magic-angle twisted bilayer graphene. *Nature* **2019**, *573*, 91–95. [CrossRef]

19. Kerelsky, A.; McGilly, L.J.; Kennes, D.M.; Xian, L.; Yankowitz, M.; Chen, S.; Watanabe, K.; Taniguchi, T.; Hone, J.; Dean, C.; et al. Maximized electron interactions at the magic angle in twisted bilayer graphene. *Nature* **2019**, *572*, 95–100. [CrossRef]

20. Xie, Y.; Lian, B.; Jack, B.; Liu, X.; Chiu, C.L.; Watanabe, K.; Taniguchi, T.; Bernevig, B.A.; Yazdani, A. Spectroscopic signatures of many-body correlations in magic-angle twisted bilayer graphene. *Nature* **2019**, *572*, 101–105. [CrossRef]

21. Burg, G.W.; Zhu, J.; Taniguchi, T.; Watanabe, K.; MacDonald, A.H.; Tutuc, E. Correlated Insulating States in Twisted Double Bilayer Graphene. *Phys. Rev. Lett.* **2019**, *123*, 197702. [CrossRef] [PubMed]

22. Zuo, W.-J.; Qiao, J.-B.; Ma, D.-L.; Yin, L.-J.; Sun, G.; Zhang, J.-Y.; Guan, L.-Y.; He, L. Scanning tunneling microscopy and spectroscopy of twisted trilayer graphene. *Phys. Rev. B* **2018**, *97*, 035440. [CrossRef]

23. Khalaf, E.; Kruchkov, A.J.; Tarnopolsky, G.; Vishwanath, A. Magic angle hierarchy in twisted graphene multilayers. *Phys. Rev. B* **2019**, *100*, 085109. [CrossRef]

24. Dos Santos, J.M.B.L.; Peres, N.M.R.; Castro Neto, A.H. Graphene bilayer with a twist: Electronic structure. *Phys. Rev. Lett.* **2007**, *99*, 256802. [CrossRef]

25. Kim, K.; Yankowitz, M.; Fallahazad, B.; Kang, S.; Movva, H.C.P.; Huang, S.; Larentis, S.; Corbet, C.M.; Taniguchi, T.; Watanabe, K.; et al. van der Waals Heterostructures with High Accuracy Rotational Alignment. *Nano Lett.* **2016**, *16*, 5968. [CrossRef]

26. Cao, Y.; Luo, J.Y.; Fatemi, V.; Fang, S.; Sanchez-Yamagishi, J.D.; Watanabe, K.; Taniguchi, T.; Kaxiras, E.; Jarillo-Herrero, P. Superlattice-Induced Insulating States and Valley-Protected Orbits in Twisted Bilayer Graphene. *Phys. Rev. Lett.* **2016**, *117*, 116804. [CrossRef]

27. Zou, L.; Po, H.C.; Vishwanath, A.; Senthi, T. Band structure of twisted bilayer graphene: Emergent symmetries, commensurate approximants, and Wannier obstructions. *Phys. Rev. B* **2018**, *98*, 085435. [CrossRef]

28. Deng, B.; Ma, C.; Wang, Q.; Yuan, S.; Watanabe, K.; Taniguchi, T.; Zhang, F.; Xia, F. Strong mid-infrared photoresponse in small-twist-angle bilayer graphene. *Nat. Photonics* **2020**, 1–5. [CrossRef]

29. Patel, H.; Havener, R.W.; Brown, L.; Liang, Y.; Yang, L.; Park, J.; Graham, M.W. Tunable Optical Excitations in Twisted Bilayer Graphene Form Strongly Bound Excitons. *Nano Lett.* **2015**, *15*, 5932–5937. [CrossRef]

30. Yin, J.; Wang, H.; Peng, H.; Tan, Z.; Liao, L.; Lin, L.; Sun, X.; Koh, A.L.; Chen, Y.; Peng, H.; et al. Selectively enhanced photocurrent generation in twisted bilayer graphene with van Hove singularity. *Nat. Commun.* **2016**, *7*, 10699. [CrossRef]

31. Patel, H.; Huang, L.; Kim, C.-J.; Park, J.; Graham, M.W. Stacking angle-tunable photoluminescence from interlayer exciton states in twisted bilayer graphene. *Nat. Commun.* **2019**, *10*, 1445. [CrossRef]

32. Yu, K.; Van Luan, N.; Kim, T.; Jeon, J.; Kim, J.; Moon, P.; Lee, Y.H.; Choi, E.J. Gate tunable optical absorption and band structure of twisted bilayer graphene. *Phys. Rev. B* **2019**, *99*, 241405. [CrossRef]

33. Po, H.C.; Watanabe, H.; Vishwanath, A. Fragile Topology and Wannier Obstructions. *Phys. Rev. Lett.* **2018**, *121*, 126402. [CrossRef] [PubMed]

34. Ahn, J.; Park, S.; Yang, B.-J. Failure of Nielsen-Ninomiya Theorem and Fragile Topology in Two-Dimensional Systems with Space-Time Inversion Symmetry: Application to Twisted Bilayer Graphene at Magic Angle. *Phys. Rev. X* **2019**, *9*, 021013. [CrossRef]

35. Hejazi, K.; Liu, C.; Shapourian, H.; Chen, X.; Balents, L. Multiple topological transitions in twisted bilayer graphene near the first magic angle. *Phys. Rev. B* **2019**, *99*, 035111. [CrossRef]

36. Liu, J.; Liu, J.; Dai, X. Pseudo Landau level representation of twisted bilayer graphene: Band topology and implications on the correlated insulating phase. *Phys. Rev. B* **2019**, *99*, 155415. [CrossRef]

37. Po, H.C.; Zou, L.; Senthil, T.; Vishwanath, A. Faithful tight-binding models and fragile topology of magic-angle bilayer graphene. *Phys. Rev. B* **2019**, *99*, 195455. [CrossRef]

38. Tarnopolsky, G.; Kruchkov, A.J.; Vishwanath, A. Origin of Magic Angles in Twisted Bilayer Graphene. *Phys. Rev. Lett.* **2019**, *122*, 106405. [CrossRef]

39. Lisi, S.; Lu, X.; Benschop, T.; Jong, T.A.d.; Stepanov, P.; Duran, J.R.; Margot, F.; Cucchi, I.; Cappelli, E.; Hunter, A.; et al. Direct evidence for flat bands in twisted bilayer graphene from nano-ARPES. *arXiv* **2020**, arXiv:2002.02289.

40. Brihuega, I.; Mallet, P.; Gonzalez-Herrero, H.; de laissardiere, G.T.; Ugeda, M.M.; Magaud, L.; Gomez-Rodriguez, J.M.; Yndurain, F.; Veuillen, J.Y. Unraveling the Intrinsic and Robust Nature of van Hove Singularities in Twisted Bilayer Graphene by Scanning Tunneling Microscopy and Theoretical Analysis. *Phys. Rev. Lett.* **2012**, *109*, 196802. [CrossRef]

41. Miller, D.L.; Kubista, K.D.; Rutter, G.M.; Ruan, M.; de Heer, W.A.; First, P.N.; Stroscio, J.A. Observing the Quantization of Zero Mass Carriers in Graphene. *Science* **2009**, *324*, 924–927. [CrossRef] [PubMed]

42. Codecido, E.; Wang, Q.; Koester, R.; Che, S.; Tian, H.; Lv, R.; Tran, S.; Watanabe, K.; Taniguchi, T.; Zhang, F.; et al. Correlated insulating and superconducting states in twisted bilayer graphene below the magic angle. *Sci. Adv.* **2019**, *5*, eaaw9770. [CrossRef]

43. Saito, Y.; Ge, J.; Watanabe, K.; Taniguchi, T.; Young, A.F. Decoupling superconductivity and correlated insulators in twisted bilayer graphene. *arXiv* **2020**, arXiv:1911.13302.

44. Liao, L.; Lin, Y.-C.; Bao, M.; Cheng, R.; Bai, J.; Liu, Y.; Qu, Y.; Wang, K.L.; Huang, Y.; Duan, X. High-speed graphene transistors with a self-aligned nanowire gate. *Nature* **2010**, *467*, 305–308. [CrossRef]

45. Wang, S.; Ang, P.K.; Wang, Z.; Tang, A.L.L.; Thong, J.T.L.; Loh, K.P. High Mobility, Printable, and Solution-Processed Graphene Electronics. *Nano Lett.* **2010**, *10*, 92–98. [CrossRef]

46. Banszerus, L.; Schmitz, M.; Engels, S.; Dauber, J.; Oellers, M.; Haupt, F.; Watanabe, K.; Taniguchi, T.; Beschoten, B.; Stampfer, C. Ultrahigh-mobility graphene devices from chemical vapor deposition on reusable copper. *Sci. Adv.* **2015**, *1*, e1500222. [CrossRef] [PubMed]

47. Liu, C.-C.; Zhang, L.-D.; Chen, W.-Q.; Yang, F. Chiral Spin Density Wave and d plus id Superconductivity in the Magic-Angle-Twisted Bilayer Graphene. *Phys. Rev. Lett.* **2018**, *121*, 217001. [CrossRef] [PubMed]

48. Wu, F.; MacDonald, A.H.; Martin, I. Theory of Phonon-Mediated Superconductivity in Twisted Bilayer Graphene. *Phys. Rev. Lett.* **2018**, *121*, 257001. [CrossRef]

49. Xu, C.; Balents, L. Topological Superconductivity in Twisted Multilayer Graphene. *Phys. Rev. Lett.* **2018**, *121*, 087001. [CrossRef]

50. Gonzalez, J.; Stauber, T. Kohn-Luttinger Superconductivity in Twisted Bilayer Graphene. *Phys. Rev. Lett.* **2019**, *122*, 026801. [CrossRef]

51. Lian, B.; Wang, Z.; Bernevig, B.A. Twisted Bilayer Graphene: A Phonon-Driven Superconductor. *Phys. Rev. Lett.* **2019**, *122*, 257002. [CrossRef]

52. Isobe, H.; Yuan, N.F.Q.; Fu, L. Unconventional Superconductivity and Density Waves in Twisted Bilayer Graphene. *Phys. Rev. X* **2018**, *8*, 041041. [CrossRef]

53. Kang, J.; Vafek, O. Symmetry, Maximally Localized Wannier States, and a Low-Energy Model for Twisted Bilayer Graphene Narrow Bands. *Phys. Rev. X* **2018**, *8*, 031088. [CrossRef]

54. Koshino, M.; Yuan, N.F.Q.; Koretsune, T.; Ochi, M.; Kuroki, K.; Fu, L. Maximally Localized Wannier Orbitals and the Extended Hubbard Model for Twisted Bilayer Graphene. *Phys. Rev. X* **2018**, *8*, 031087. [CrossRef]

55. Po, H.C.; Zou, L.; Vishwanath, A.; Senthil, T. Origin of Mott Insulating Behavior and Superconductivity in Twisted Bilayer Graphene. *Phys. Rev. X* **2018**, *8*, 031089. [CrossRef]

56. Yankowitz, M.; Chen, S.; Polshyn, H.; Zhang, Y.; Watanabe, K.; Taniguchi, T.; Graf, D.; Young, A.F.; Dean, C.R. Tuning superconductivity in twisted bilayer graphene. *Science* **2019**, *363*, 1059–1064. [CrossRef] [PubMed]

57. Balents, L.; Dean, C.R.; Efetov, D.K.; Young, A.F. Superconductivity and strong correlations in moiré flat bands. *Nat. Phys.* **2020**, 1–9. [CrossRef]

58. Wu, F.; Das Sarma, S. Collective Excitations of Quantum Anomalous Hall Ferromagnets in Twisted Bilayer Graphene. *Phys. Rev. Lett.* **2020**, *124*, 046403. [CrossRef] [PubMed]

59. Xie, M.; MacDonald, A.H. Nature of the Correlated Insulator States in Twisted Bilayer Graphene. *Phys. Rev. Lett.* **2020**, *124*, 097601. [CrossRef] [PubMed]

60. Yankowitz, M.; Jung, J.; Laksono, E.; Leconte, N.; Chittari, B.L.; Watanabe, K.; Taniguchi, T.; Adam, S.; Graf, D.; Dean, C.R. Dynamic band-structure tuning of graphene moiré superlattices with pressure. *Nature* **2018**, *557*, 404–408. [CrossRef]

61. Mogera, U.; Kulkarni, G.U. A new twist in graphene research: Twisted graphene. *Carbon* **2020**, *156*, 470–487. [CrossRef]

62. Gonzalez-Arraga, L.A.; Lado, J.L.; Guinea, F.; San-Jose, P. Electrically Controllable Magnetism in Twisted Bilayer Graphene. *Phys. Rev. Lett.* **2017**, *119*, 107201. [CrossRef]

63. Thomson, A.; Chatterjee, S.; Sachdev, S.; Scheurer, M.S. Triangular antiferromagnetism on the honeycomb lattice of twisted bilayer graphene. *Phys. Rev. B* **2018**, *98*, 075109. [CrossRef]

64. Sharpe, A.L.; Fox, E.J.; Barnard, A.W.; Finney, J.; Watanabe, K.; Taniguchi, T.; Kastner, M.A.; Goldhaber-Gordon, D. Emergent ferromagnetism near three-quarters filling in twisted bilayer graphene. *Science* **2019**, *365*, 605–608. [CrossRef] [PubMed]

65. Liu, J.; Ma, Z.; Gao, J.; Dai, X. Quantum Valley Hall Effect, Orbital Magnetism, and Anomalous Hall Effect in Twisted Multilayer Graphene Systems. *Phys. Rev. X* **2019**, *9*, 031021. [CrossRef]

66. Bultinck, N.; Chatterjee, S.; Zaletel, M.P. Mechanism for Anomalous Hall Ferromagnetism in Twisted Bilayer Graphene. *Phys. Rev. Lett.* **2020**, *124*, 166601. [CrossRef] [PubMed]

67. He, W.-Y.; Goldhaber-Gordon, D.; Law, K.T. Giant orbital magnetoelectric effect and current-induced magnetization switching in twisted bilayer graphene. *Nat. Commun.* **2020**, *11*, 1650. [CrossRef]

68. Repellin, C.; Senthil, T. Chern bands of twisted bilayer graphene: Fractional Chern insulators and spin phase transition. *Phys. Rev. Res.* **2020**, *2*, 023238. [CrossRef]

69. Chittari, B.L.; Chen, G.; Zhang, Y.; Wang, F.; Jung, J. Gate-Tunable Topological Flat Bands in Trilayer Graphene Boron-Nitride Moiré Superlattices. *Phys. Rev. Lett.* **2019**, *122*, 016401. [CrossRef]

70. Yin, L.-J.; Jiang, H.; Qiao, J.-B.; He, L. Direct imaging of topological edge states at a bilayer graphene domain wall. *Nat. Commun.* **2016**, *7*, 11760. [CrossRef]

71. Huang, S.; Kim, K.; Efimkin, D.K.; Lovorn, T.; Taniguchi, T.; Watanabe, K.; MacDonald, A.H.; Tutuc, E.; LeRoy, B.J. Topologically Protected Helical States in Minimally Twisted Bilayer Graphene. *Phys. Rev. Lett.* **2018**, *121*, 037702. [CrossRef]

72. Uri, A.; Grover, S.; Cao, Y.; Crosse, J.A.; Bagani, K.; Rodan-Legrain, D.; Myasoedov, Y.; Watanabe, K.; Taniguchi, T.; Moon, P.; et al. Mapping the twist-angle disorder and Landau levels in magic-angle graphene. *Nature* **2020**, *581*, 47–52. [CrossRef] [PubMed]

73. Vasyukov, D.; Anahory, Y.; Embon, L.; Halbertal, D.; Cuppens, J.; Neeman, L.; Finkler, A.; Segev, Y.; Myasoedov, Y.; Rappaport, M.L.; et al. A scanning superconducting quantum interference device with single electron spin sensitivity. *Nat. Nanotechnol.* **2013**, *8*, 639–644. [CrossRef] [PubMed]

74. Shen, C.; Chu, Y.; Wu, Q.; Li, N.; Wang, S.; Zhao, Y.; Tang, J.; Liu, J.; Tian, J.; Watanabe, K.; et al. Correlated states in twisted double bilayer graphene. *Nat. Phys.* **2020**, *16*, 520–525. [CrossRef]

75. Cao, Y.; Rodan-Legrain, D.; Rubies-Bigorda, O.; Park, J.M.; Watanabe, K.; Taniguchi, T.; Jarillo-Herrero, P. Tunable correlated states and spin-polarized phases in twisted bilayer–bilayer graphene. *Nature* **2020**, 1–6. [CrossRef]

76. He, M.; Li, Y.; Cai, J.; Liu, Y.; Watanabe, K.; Taniguchi, T.; Xu, X.; Yankowitz, M. Tunable correlation-driven symmetry breaking in twisted double bilayer graphene. *arXiv* **2020**, arXiv:2002.08904.

77. Liu, X.; Hao, Z.; Khalaf, E.; Lee, J.Y.; Watanabe, K.; Taniguchi, T.; Vishwanath, A.; Kim, P. Spin-polarized Correlated Insulator and Superconductor in Twisted Double Bilayer Graphene a. *arXiv* **2019**, arXiv:1903.08130.

78. Lee, J.Y.; Khalaf, E.; Liu, S.; Liu, X.; Hao, Z.; Kim, P.; Vishwanath, A. Theory of correlated insulating behaviour and spin-triplet superconductivity in twisted double bilayer graphene. *Nat. Commun.* **2019**, *10*, 5333. [CrossRef] [PubMed]

79. Chen, G.; Jiang, L.; Wu, S.; Lyu, B.; Li, H.; Chittari, B.L.; Watanabe, K.; Taniguchi, T.; Shi, Z.; Jung, J.; et al. Evidence of a gate-tunable Mott insulator in a trilayer graphene moiré superlattice. *Nat. Phys.* **2019**, *15*, 237–241. [CrossRef]

80. Chen, G.; Sharpe, A.L.; Gallagher, P.; Rosen, I.T.; Fox, E.J.; Jiang, L.; Lyu, B.; Li, H.; Watanabe, K.; Taniguchi, T.; et al. Signatures of tunable superconductivity in a trilayer graphene moiré superlattice. *Nature* **2019**, *572*, 215–219. [CrossRef]

81. Yin, L.-J.; Shi, L.-J.; Li, S.-Y.; Zhang, Y.; Guo, Z.-H.; He, L. High-Magnetic-Field Tunneling Spectra of ABC-Stacked Trilayer Graphene on Graphite. *Phys. Rev. Lett.* **2019**, *122*, 146802. [CrossRef] [PubMed]

82. Chen, S.; He, M.; Zhang, Y.-H.; Hsieh, V.; Fei, Z.; Watanabe, K.; Taniguchi, T.; Cobden, D.H.; Xu, X.; Dean, C.R.; et al. Electrically tunable correlated and topological states in twisted monolayer-bilayer graphene. *arXiv* **2020**, arXiv:2004.11340.

83. Gupta, N.; Walia, S.; Mogera, U.; Kulkarni, G.U. Twist-Dependent Raman and Electron Diffraction Correlations in Twisted Multilayer Graphene. *J. Phys. Chem. Lett.* **2020**, *11*, 2797–2803. [CrossRef] [PubMed]

84. Cea, T.; Walet, N.R.; Guinea, F. Twists and the Electronic Structure of Graphitic Materials. *Nano Lett.* **2019**, *19*, 8683–8689. [CrossRef]

85. Carr, S.; Li, C.; Zhu, Z.; Kaxiras, E.; Sachdev, S.; Kruchkov, A. Coexistence of ultraheavy and ultrarelativistic Dirac quasiparticles in sandwiched trilayer graphene. *arXiv* **2019**, arXiv:1907.00952.

86. Ma, Z.; Li, S.; Zheng, Y.-W.; Xiao, M.-M.; Jiang, H.; Gao, J.-H.; Xie, X.C. Topological flat bands in twisted trilayer graphene. *arXiv* **2019**, arXiv:1905.00622.

87. Castellanos-Gomez, A.; Buscema, M.; Molenaar, R.; Singh, V.; Janssen, L.; van der Zant, H.S.J.; Steele, G.A. Deterministic transfer of two-dimensional materials by all-dry viscoelastic stamping. *2D Mater.* **2014**, *1*, 011002. [CrossRef]

88. Chen, X.-D.; Xin, W.; Jiang, W.-S.; Liu, Z.-B.; Chen, Y.; Tian, J.-G. High-Precision Twist-Controlled Bilayer and Trilayer Graphene. *Adv. Mate.* **2016**, *28*, 2563. [CrossRef]

89. Hass, J.; Varchon, F.; Millán-Otoya, J.E.; Sprinkle, M.; Sharma, N.; de Heer, W.A.; Berger, C.; First, P.N.; Magaud, L.; Conrad, E.H. Why Multilayer Graphene on 4H-SiC(000-1) Behaves Like a Single Sheet of Graphene. *Phys. Rev. Lett.* **2008**, *100*, 125504. [CrossRef]

90. De Heer, W.A.; Berger, C.; Wu, X.; First, P.N.; Conrad, E.H.; Li, X.; Li, T.; Sprinkle, M.; Hass, J.; Sadowski, M.L.; et al. Epitaxial graphene. *Solid State Commun.* **2007**, *143*, 92–100. [CrossRef]

91. Emtsev, K.V.; Bostwick, A.; Horn, K.; Jobst, J.; Kellogg, G.L.; Ley, L.; McChesney, J.L.; Ohta, T.; Reshanov, S.A.; Rohrl, J.; et al. Towards wafer-size graphene layers by atmospheric pressure graphitization of silicon carbide. *Nat. Mater.* **2009**, *8*, 203–207. [CrossRef] [PubMed]

92. Cherkez, V.; de Laissardiere, G.T.; Mallet, P.; Veuillen, J.Y. Van Hove singularities in doped twisted graphene bilayers studied by scanning tunneling spectroscopy. *Phys. Rev. B* **2015**, *91*, 155428. [CrossRef]

93. Van Bommel, A.J.; Crombeen, J.E.; Van Tooren, A. LEED and Auger electron observations of the SiC(0001) surface. *Surf. Sci.* **1975**, *48*, 463–472. [CrossRef]

94. Naitoh, M.; Kitada, M.; Nishigaki, S.; Toyama, N.; Shoji, F. An STM Observation of the Initial Process of Graphitization at the 6H-SiC(000-1) surface. *Surf. Rev. Lett.* **2003**, *10*, 473–477. [CrossRef]

95. Berger, C.; Song, Z.M.; Li, T.B.; Li, X.B.; Ogbazghi, A.Y.; Feng, R.; Dai, Z.T.; Marchenkov, A.N.; Conrad, E.H.; First, P.N.; et al. Ultrathin epitaxial graphite: 2D electron gas properties and a route toward graphene-based nanoelectronics. *J. Phys. Chem. B* **2004**, *108*, 19912–19916. [CrossRef]

96. Berger, C.; Song, Z.; Li, X.; Wu, X.; Brown, N.; Naud, C.; Mayou, D.; Li, T.; Hass, J.; Marchenkov, A.N.; et al. Electronic Confinement and Coherence in Patterned Epitaxial Graphene. *Science* **2006**, *312*, 1191–1196. [CrossRef]

97. Ong, W.J.; Tok, E.S. Role of Si clusters in the phase transformation and formation of (6×6)-ring structures on 6H-SiC(0001) as a function of temperature: An STM and XPS study. *Phys. Rev. B* **2006**, *73*, 045330. [CrossRef]

98. Rollings, E.; Gweon, G.H.; Zhou, S.Y.; Mun, B.S.; McChesney, J.L.; Hussain, B.S.; Fedorov, A.V.; First, P.N.; de Heer, W.A.; Lanzara, A. Synthesis and characterization of atomically thin graphite films on a silicon carbide substrate. *J. Phys. Chem. Solids* **2006**, *67*, 2172–2177. [CrossRef]

99. Seyller, T.; Emtsev, K.V.; Gao, K.; Speck, F.; Ley, L.; Tadich, A.; Broekman, L.; Riley, J.D.; Leckey, R.C.G.; Rader, O.; et al. Structural and electronic properties of graphite layers grown on SiC(0001). *Surf. Sci.* **2006**, *600*, 3906–3911. [CrossRef]

100. Hibino, H.; Kageshima, H.; Maeda, F.; Nagase, M.; Kobayashi, Y.; Yamaguchi, H. Microscopic thickness determination of thin graphite films formed on SiC from quantized oscillation in reflectivity of low-energy electrons. *Phys. Rev. B* **2008**, *77*, 075413. [CrossRef]

101. Hiebel, F.; Mallet, P.; Varchon, F.; Magaud, L.; Veuillen, J.Y. Graphene-substrate interaction on 6H-SiC(000(1)over bar): A scanning tunneling microscopy study. *Phys. Rev. B* **2008**, *78*, 153412. [CrossRef]

102. Ivanov, I.G.; Hassan, J.U.; Iakimov, T.; Zakharov, A.A.; Yakimova, R.; Janzen, E. Layer-Number determination in graphene on SiC by reflectance mapping. *Carbon* **2014**, *77*, 492–500. [CrossRef]

103. Tedesco, J.L.; Jernigan, G.G.; Culbertson, J.C.; Hite, J.K.; Yang, Y.; Daniels, K.M.; Myers-Ward, R.L.; Eddy, C.R.; Robinson, J.A.; Trumbull, K.A.; et al. Morphology characterization of argon-mediated epitaxial graphene on C-face SiC. *Appl. Phys. Lett.* **2010**, *96*, 222103. [CrossRef]

104. Yazdi, G.R.; Vasiliauskas, R.; Iakimov, T.; Zakharov, A.; Syväjärvi, M.; Yakimova, R. Growth of large area monolayer graphene on 3C-SiC and a comparison with other SiC polytypes. *Carbon* **2013**, *57*, 477–484. [CrossRef]

105. Wang, W.-X.; Jiang, H.; Zhang, Y.; Li, S.-Y.; Liu, H.; Li, X.; Wu, X.; He, L. Scanning tunneling microscopy and spectroscopy of finite-size twisted bilayer graphene. *Phys. Rev. B* **2017**, *96*, 115434. [CrossRef]

106. Marchenko, D.; Evtushinsky, D.V.; Golias, E.; Varykhalov, A.; Seyller, T.; Rader, O. Extremely flat band in bilayer graphene. *Sci. Adv.* **2018**, *4*, 7. [CrossRef]

107. Wang, W.; Shi, Y.; Zakharov, A.A.; Syvajarvi, M.; Yakimova, R.; Uhrberg, R.I.G.; Sun, J. Flat-Band Electronic Structure and Interlayer Spacing Influence in Rhombohedral Four-Layer Graphene. *Nano Lett.* **2018**, *18*, 5862–5866. [CrossRef]